Prepared by the Special Publications Division, National Geographic Society, Washington, D. C.

The National Geographic Society extends its appreciation to the Ocean Trust Foundation, a nonprofit scientific organization, and to PolyGram Pictures for their cooperation in the preparation of this book.

Exploring the Deep Frontier

The Adventure of Man in the Sea

by Sylvia A. Earle and Al Giddings

GEOGRAPHIC SOCIETY

EXPLORING THE DEEP FRONTIER:
THE ADVENTURE OF MAN IN THE SEA

Published by
The National Geographic Society
ROBERT E. DOYLE, *President*
MELVIN M. PAYNE, *Chairman of the Board*
GILBERT M. GROSVENOR, *Editor*
MELVILLE BELL GROSVENOR, *Editor Emeritus*
Prepared by
The Special Publications Division
ROBERT L. BREEDEN, *Editor*
DONALD J. CRUMP, *Associate Editor*
PHILIP B. SILCOTT, *Senior Editor*
JOAN TAPPER, *Managing Editor*
WILLIAM R. GRAY, *Consulting Editor*
THOMAS B. POWELL III, *Picture Editor*
JODY BOLT, *Art Director*
BONNIE S. LAWRENCE, *Senior Researcher and Assistant to the Editor*
STEPHEN HUBBARD, KAREN M. KOSTYAL,
LOUISA MAGZANIAN, PENNY DIAMANTI DE WIDT,
KATHLEEN F. TETER, *Researchers;*
TONI EUGENE, BARBARA GRAZZINI,
ANNMARIE MANZI, *Additional Research Assistance*
Illustrations and Design
SUEZ B. KEHL, *Assistant Art Director*
CINDA ROSE, MARIANNE RIGLER KOSZORUS, *Assistant Designers*
KENNETH R. SHRADER, CYNTHIA BREEDEN, *Design Assistants*
LOUIS DE LA HABA, TONI EUGENE, RON FISHER,
CHRISTINE ECKSTROM LEE, TOM MELHAM,
H. ROBERT MORRISON, *Picture Legends*
JOHN D. GARST, JR., GARY M. JOHNSON, MARK SEIDLER,
Map Research, Design, and Production
Engraving, Printing, and Product Manufacture
ROBERT W. MESSER, *Manager*
GEORGE V. WHITE, *Production Manager*
JUNE L. GRAHAM, RICHARD A. MCCLURE,
RAJA D. MURSHED, CHRISTINE A. ROBERTS,
Assistant Production Managers
DAVID V. SHOWERS, *Production Assistant*
SUSAN M. OEHLER, *Production Staff Assistant*
DEBRA A. ANTONINI, PAMELA A. BLACK, BARBARA BRICKS,
JANE H. BUXTON, MARY ELIZABETH DAVIS,
ROSAMUND GARNER, NANCY J. HARVEY, JANE M. HOLLOWAY,
SUZANNE J. JACOBSON, ARTEMIS S. LAMPATHAKIS,
CLEO PETROFF, MARCIA ROBINSON, KATHERYN M. SLOCUM,
SUZANNE VENINO, *Staff Assistants*
JEFFREY A. BROWN, ELIZABETH MEYENDORFF, *Index*

Library of Congress CIP Data: page 295

Riding the platform of the flag-bedecked submersible Star II, *author Sylvia A. Earle, protected by a pressurized diving suit, prepares to step onto the seafloor during a test dive near Oahu, Hawaii.* PRECEDING PAGES: *Labyrinth of stony coral beckons a scuba diver to a watery world. Page 1: Parrot-beaked blue angelfish swims among coral branches off Bermuda. Endpapers: Divers explore an undersea canyon in Australia's Great Barrier Reef. Hardcover: Surrounded by spadefish, two scuba divers probe the depths.*

CHUCK NICKLIN / SEA FILMS, INC. (LEFT); DAVID DOUBILET (PRECEDING PAGES AND PAGE 1); N.G.S. PHOTOGRAPHER BATES LITTLEHALES (ENDPAPERS)

Contents

JERRY GREENBERG

Bubbles streaming from their diving gear, two teams of aquanauts change places. The submersible Deep Diver *shuttles the crews between Igloo, an inflatable rubber underwater workroom, and the surface.*

Our Planet's Life Belt

Jacques-Yves Cousteau

I love the sea, born of an inconceivable deluge, because it is made of water. Water, as fluid as our souls, shapeless, enslaved to none but to gravity. Water, welcoming our bodies in a total embrace, setting us free from our weight. Water, mother of all life, fragile guarantor of our survival.

The ocean has roots—the rivers, which draw minerals and nutrients from the hearts of the continents. In the bosom of the sea multitudes of creatures were conceived; they evolved, they invaded the land, and some of them returned to aquatic life. The sea level rose and fell with the rhythm of glaciations, leaving scars on the continents and deep fossil beaches. Partner of the sun, the ocean is the boiler and condenser of a gigantic steam engine, as powerful as millions of nuclear plants, governing crops, floods, droughts, frosts, hurricanes, at times as serene as motherhood or as frightening as God's wrath.

I love the sea, which I have blindly trusted since the day I dared become a "manfish." Since then, in thousands of dives, I have become increasingly eager to discover the unknown facets of the ocean world. After half a century of that tumultuous love affair, it is time for me to measure the progress made, to assess the damage done, and to answer questions raised by my conscience.

During the past 50 years world fisheries have, in my opinion, made little progress, despite a vast increase in the tonnage of fishing fleets, despite the introduction of nylon nets and sonar, despite the dissemination of scientific data. In fact, the catches of some traditionally commercial fishes are on the decline; global figures are artificially maintained by fishing new places and taking species hitherto neglected. The decreasing size of fishes and crustaceans harvested suggests that the seas are overfished. We are spending our capital, not our income. If there were effective world fisheries management, emergency measures would have been taken 10 years ago.

Meanwhile, as much as 30 percent of the fishing crop goes to feed livestock and poultry, while, it seems to me, a hungry, expanding human population obtains less protein from the sea. Aquaculture—the cultivation of marine organisms—has yet to reach its potential. Only shellfish have been actively developed. But shrimp and lobsters are not what we need to feed starving people. We need hundreds of millions of tons of inexpensive fish, and we still have no idea how to obtain such a yield.

In the 1950s great mineral riches were discovered on the seafloor. Yet 30 years later, in spite of impressive marine technologies, these finds remain virtually unused and unmanaged. The picture is equally grim if we assess what has been done with the vast quantity of energy dispensed into the sea. The possibilities of exploiting tide, salinity gradient, ocean thermal energy, wind, bioconversion—all have been scanned timidly and declared unrealistic, although the tidal energy plant at La Rance, France, has run since 1966, ocean thermal energy conversion was tested in 1929, and windmills have helped to drain Dutch polders for centuries.

Many of the large corporations that invested in ocean research and development around 1960 abandoned their efforts 10 years later, estimating that the profitable years were still far ahead and that government should prime the financial pump. Why such disenchantment? Why such passivity? While it is true that resources exist in the sea, they are scattered, difficult to reach, and vulnerable to pollution and careless exploitation. For the vast majority of human beings, moreover, the ocean remains a romantic subject, a setting for a pleasure cruise, not a resource to be managed.

Paradoxically, while decision-makers hesitate and living ocean resources dwindle, the public is awakening to the existence of sea resources and to the man-made dangers that threaten them.

Science and technology have already stored up enough knowledge to monitor the health of the ocean from satellites; to extract mineral and living resources without depleting them; to protect our lakes, rivers, seas, and polar ice caps; to manage the populations of marine mammals; to avoid the decay of coral reefs. But we still must convince the leaders of the world that there is only one water system on our planet, that its protection can be achieved only by applying the same set of guidelines to all nations. Most important, those leaders must understand that the very survival of the human species depends upon the maintenance of an ocean clean and alive, spreading all around the world. The ocean is our planet's life belt.

Undersea pioneer Capt. Jacques-Yves Cousteau emerges from his diving saucer, a research submersible that has plumbed waters worldwide.

N.G.S. PHOTOGRAPHER BATES LITTLEHALES

Prologue: Water, Essence of Life

Aglitter with sunlight, a living tunnel forms around a diver as a school of herring swirls past in waters off Cozumel, an island near Mexico.

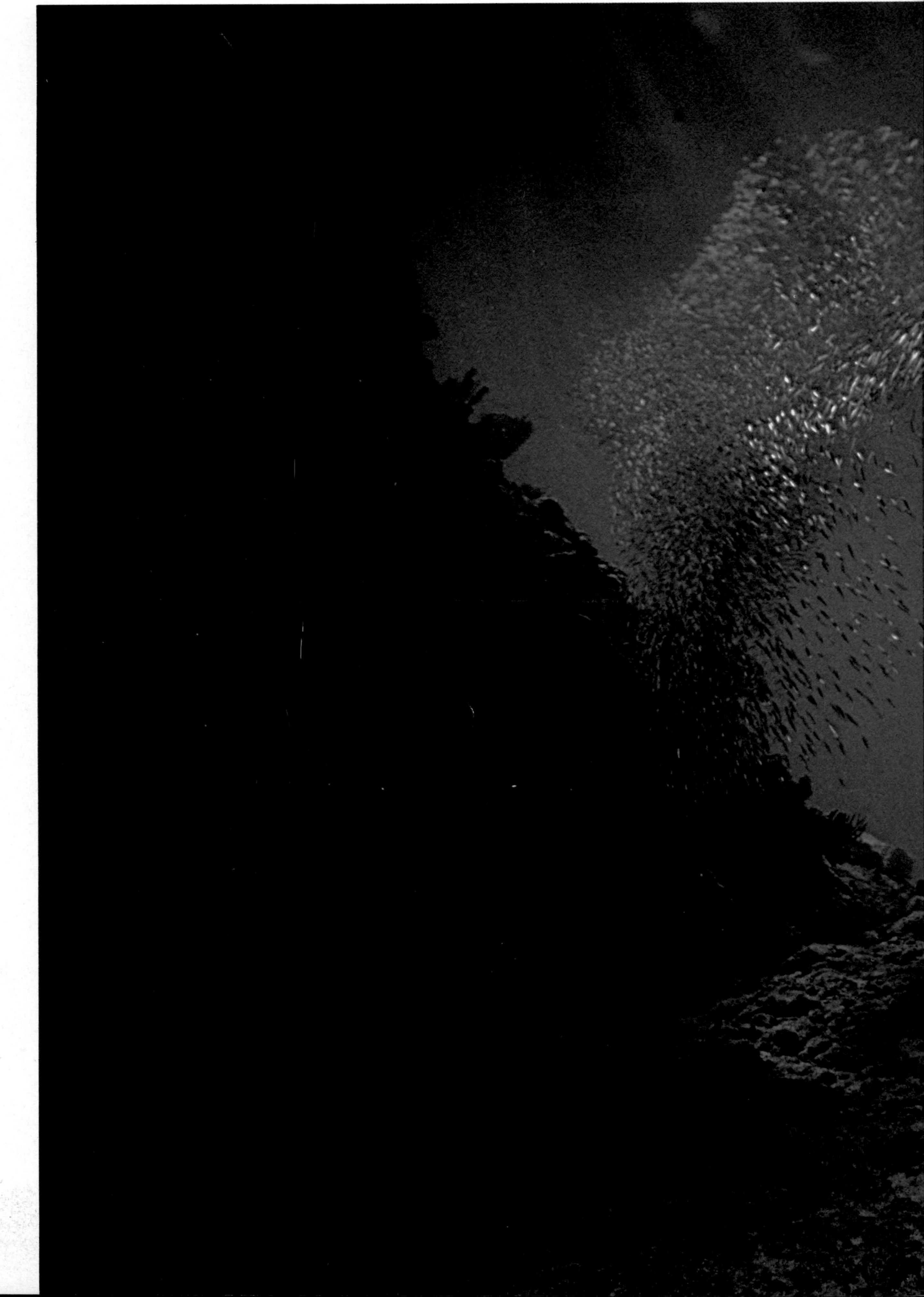

"For the sea lies all about us. . . . The continents themselves dissolve and pass to the sea. . . .
In its mysterious past it encompasses all the dim origins of life . . . and receives in the end . . .
the dead husks of that same life. For all at last return . . . to Oceanus, the ocean river,
like the everflowing stream of time, the beginning and the end." Rachel Carson, *The Sea Around Us*

GERI MURPHY

ONE HUNDRED FEET beneath a California inlet we hover, eye to eye. One of us, a 110-pound diver, carries air tanks and wears flippers, mask, and rubber suit—a land dweller briefly exploring another world through the wonders of 20th-century ingenuity. The other, an eight-ounce squid, is sleek, bare, glistening—a sea creature superbly adapted to life in a cool, salt-water atmosphere. Silently we pause and regard each other. Minutes pass; then with a smoothness that suggests no motion at all, the silvery creature glides out of sight, into a dark sea.

In 1980, as we reach into outer space, exploring, seeking to find life in other parts of our solar system and beyond, the liquid realm around us remains nearly as unprobed as the distant planets. There is good reason to marvel at how little we know about the sea. Yet a greater wonder may be that we know as much as we do, considering the obstacles that confront us: depth, pressure, cold, darkness, and inaccessibility. Methods to help overcome these barriers have been developed only recently.

Never before has a generation been better equipped, through technology, to explore the oceans. Never before has there been greater interest in discovering the physical and chemical nature of the seafloor, of the living systems that exist in the ocean, of the water itself. Never before have we so needed to know how the sea relates to our survival and well-being—how to protect those systems that yield, unbidden, what we have always taken for granted: oxygen, fresh water, food, even climate.

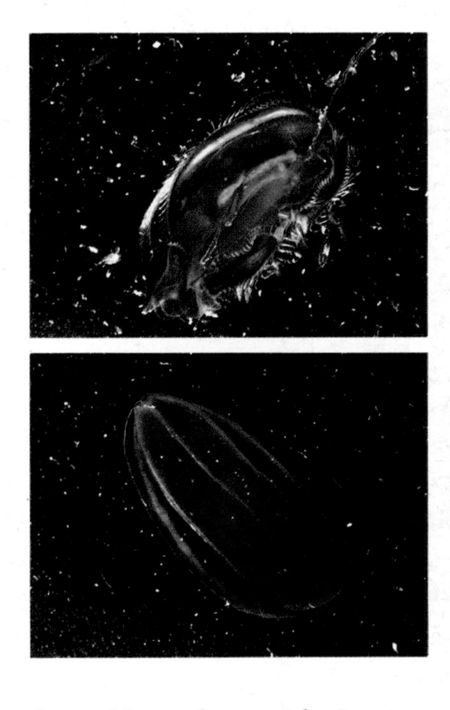

Ignited by a photographer's light, tiny marine animals brought up from beneath Arctic ice flash with color. Below, author Sylvia Earle peers at a squid and, below left, examines masses of squid eggs as part of a study of kelp forests.

AL GIDDINGS / SEA FILMS, INC. (BELOW, LEFT AND RIGHT); NATIONAL GEOGRAPHIC PHOTOGRAPHER EMORY KRISTOF

Probing the frontiers of undersea exploration, diver-scientists lower a camera into 13,000-foot Arctic Ocean depths (opposite). The eerie glow of sunlight barely penetrates the frigid waters beneath an ice floe.

STEPHEN C. EARLEY

The probing of the past few decades confirms that the earth is alone among the planets of our solar system in having abundant water in three forms: vapor, ice, and liquid. That water not only is present but also dominates our planet is obvious to astronauts from their distant perspective in space. Michael Collins observes in *Flying to the Moon and Other Strange Places:* "I will never forget how beautiful the earth appears from a great distance, floating silently and serenely like a blue and white marble against the pure black of space."

We think of the seas as covering nearly three-quarters of the earth, but the living space they afford is actually much greater. Land is inhabited down to a few dozen feet; treetops mark the uppermost reaches of nearly all but high-flying birds, certain insects, aircraft, and drifting microbes. But water—whether near-freezing Arctic sea or sun-warmed tropical lagoon, brightly illuminated open ocean or the

In the exuberant architecture of a coral reef in the Fiji Islands (opposite), many organisms find—and provide—both sustenance and shelter. Demands of life in the sea prompt unique adaptations. To survive, creatures have evolved shapes, colors, and life-styles as varied as those of the swimming crab (top, right), the banded coral shrimp (center, left), the shell-dwelling marine hermit crab (center, right), and the slipper lobster (bottom).

M. TIMOTHY O'KEEFE

GERI MURPHY

DAVID DOUBILET

CHRIS NEWBERT (OPPOSITE)

PRECEDING PAGES: In streaming sunshine, sea lions cavort near a reef in the Galapagos Islands. The bold and playful 800-pound mammals spend most of their lives in the water.

DAVID DOUBILET (PRECEDING PAGES)

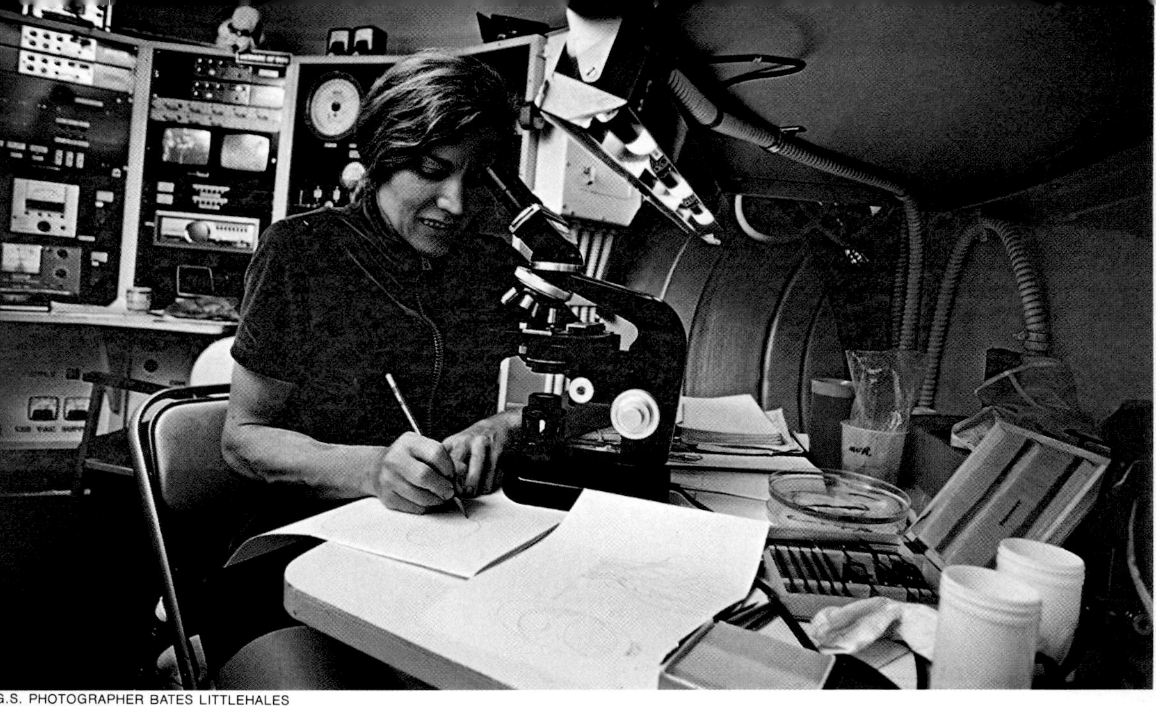

N.G.S. PHOTOGRAPHER BATES LITTLEHALES

AUTOCHROME BY W. H. LONGLEY AND CHARLES MARTIN

At home with their subjects, scientists labor to reveal the ocean's secrets. The author (top) examines microscopic sea life in an underwater habitat laboratory off the U. S. Virgin Islands. Above, a hogfish unwittingly poses for one of the first underwater color photographs ever made. Now developed to a sophisticated level, underwater color photography did not make its first published appearance until 1927. Outside a research structure in the Red Sea (opposite), divers collect marine specimens in plastic bags.

ROBERT B. GOODMAN (OPPOSITE)

great, dark depths—supports varied life from surface to seafloor.

To some, the abundance and, perhaps more significantly, the diversity of aquatic life are surprising. Of the roughly one million species of animal life, about 4 percent are those most familiar to man—vertebrates. Some three-quarters are insects. Nearly all the others are aquatic. Of the major divisions of animals and plants, each has at least some marine representation, and many live largely or entirely in the sea.

I am consistently amazed at the individuality of each member of each species. Just as we know that no two people are ever exactly alike, not even identical twins, we know that this is also true of cats and dogs, whales, fishes, and even hermit crabs.

Almost all living things share certain basic needs: space, oxygen, nutrients, a way to reproduce, appropriate temperatures within a narrow range. Our awareness that these necessities are common to most is relatively new, as is the perception that our well-being is linked to that of other life forms. Once it was thought that the sea was infinite—or at least so vast that nothing we could do would alter its nature. The riches of the sea—fishes, shrimps, whales—were regarded as inexhaustible, free for the taking, an attitude reflected in an old Yarmouth sea chanty:

"The farmer has his rent to pay.
Haul, you joskins, haul.
And seed to buy, I've heard him say.
Haul, you joskins, haul.
But we who plough the North Sea deep
Though never sowing, always reap,
The harvest which to all is free....
Haul, you joskins, haul."

We now know that nothing is truly free, that taking from one place means a trade-off somewhere else. The ocean, as earth's life-support system, already may be yielding its most significant wealth by providing basic ingredients necessary for life, things that mankind has taken for granted. As President John F. Kennedy observed in 1961, "Knowledge of the oceans is more than a matter of curiosity. Our very survival may hinge upon it."

In the past century, during the time of the greatest man-induced changes, more has been learned about the sea than during all preceding history. This era of discovery began with the scientific findings of H.M.S. *Challenger,* the first oceanographic vessel to probe

the waters of the world. The great strides in man's personal adventures in the sea are even more recent than that. In 1967, in the foreword to the Special Publication *World Beneath the Sea,* NATIONAL GEOGRAPHIC Editor Gilbert M. Grosvenor stressed this point: "For centuries man probed the mysteries of the sea, venturing upon it in frail ships or exploring its shallows in crude diving machines. But only in the past three decades has he begun to study the depths with his own eyes. From coral reefs, and even down to the dark abysses, he is uncovering secrets nature has guarded for millions of years."

In the present volume Al Giddings—along with many other talented underwater photographers—and I have presented some of the drama and excitement of these years of discovery. Several chapters describe why man has entered the ocean—the lures of curiosity, resources, treasure, and military advantage that have drawn people into the depths. In other sections we follow the methods used to probe the watery realm, from free diving to ingenious underwater craft to habitats that turn human beings into aquanauts.

Technology—sometimes considered a spoiler of civilization—has proved essential in providing prolonged access to the sea. A few free divers may be able to descend more than 300 feet underwater, but most of us, without some technological aid, cannot plunge deeper

In a blue dreamworld, new and imaginative tools extend the range of tasks human divers can perform. For transport and shelter, engineers have designed submersibles like the PC-1202 (opposite), capable of taking two crew members to 1,200 feet. The vehicle can remain underwater for eight hours, with divers entering and leaving through an air lock. Above, U. S. Navy divers test a power saw on a piece of scrap metal and, at right, set up devices to measure subsurface currents.

PERRY OCEANOGRAPHICS, INC. (OPPOSITE)

BERNIE CAMPOLI / U. S. NAVY

ROBERT B. COLE / U. S. NAVY

Startling portrait of earth, based upon thousands of depth soundings, reveals a realm that tests the imagination. Ridges and fractures, soaring peaks and flat abyssal plains distinguish the landscape beneath the oceans. The Mid-Oceanic Ridge, an underwater mountain chain, extends for 40,000 miles.

than the deep end of a swimming pool. How much beauty, how many vital aspects of life we would miss were it not for Aqua-Lungs and rebreathers, underwater laboratories and submersibles.

Not overlooked are the problems of recovering the ocean's wealth—oil, metals, food—while trying to retain its basic good health. And finally there is a look ahead to the future of the deep frontier: diving in space-age suits, under extreme conditions, with the aid of unmanned "eyes beneath the sea."

The pioneers of ocean exploration weave through these pages as they have through the history of underwater discovery in this century: naturalist Dr. William Beebe; Capt. Jacques-Yves Cousteau,

PAINTINGS BY HEINRICH C. BERANN BASED ON BATHYMETRIC STUDIES BY BRUCE C. HEEZEN AND MARIE THARP OF THE LAMONT-DOHERTY GEOLOGICAL OBSERVATORY

co-inventor of the Aqua-Lung; Dr. Harold E. Edgerton, designer of many camera systems used in the ocean; Edwin A. Link, inventor and undersea explorer; archaeologist Dr. George F. Bass; Capt. George Bond, pioneer of saturation diving.

An encyclopedic account would be impossible to present here; instead, a few significant highlights convey some of the excitement, pleasure, and wonder that await those who venture into the ocean. The invitation extended here—to enjoy the sea by becoming part of it—has been expressed delightfully by poet A. M. Sullivan:

"Plumb the bottom of the sea?
The merman buckles on his flasks
Lifts the ocean's lid and asks,
'Which of you will follow me?' "

The Lure of the Sea

WHY DIVE? What could entice a perfectly normal air-breathing human being to abandon terra firma and step into the liquid atmosphere that engulfs nearly three-quarters of the world? Curiosity? Food? The strategic advantages of warfare? Lost treasure? The simple pleasure of feeling weightless in another dimension?

For me diving was easy. In my childhood I used "swimming pools" on a grand scale: first the waters along the New Jersey coast and later those of the Gulf of Mexico. Man-made pools tend to be antiseptic places, bounded by bare walls. But mine were alive with miraculous things that teased my imagination and lured me into the sea: soft white sea anemones, golden-scaled fishes, iridescent creatures that resembled nothing I knew on land, except perhaps soap bubbles or rainbows. When I first used an Aqua-Lung in 1953, I was taking a class in marine biology at the Alligator Harbor Laboratory of Florida State University; my professor, Dr. Harold J. Humm, made it clear that he thought the best way to study fishes was to go where the fishes were, to meet them on their own terms. I did, and my life was forever changed.

The first recorded scientific underwater expedition was the mutual adventure of a distinguished zoology professor at the Sorbonne, Dr. Henri Milne-Edwards, and a naturalist friend. Instead of relying for their scientific studies on mangled creatures dragged to the surface in a net, they planned to "pursue marine animals into their most hidden retreats," using a diving helmet supplied with compressed air pumped from the surface.

The year was 1844; the place, the Strait of Messina off Italy. There, 135 years later, I too immersed myself in the cool swift waters near the legendary rock Scylla and the whirlpool Charybdis. I thought then of the 44-year-old Milne-Edwards, jumping into the unknown sea with a cumbersome helmet strapped to his shoulders. What could cause an obviously intelligent man to dive into an unfamiliar ocean, with equipment designed for the Paris Fire Brigade to use in flooded cellars? I laughed, easily imagining what moved Milne-Edwards to take the plunge: soft white sea anemones, golden-scaled fishes, iridescent creatures . . .

Milne-Edwards was but one of many sea-oriented scientists at the Sorbonne. Another, Louis Boutan, carved a niche in history in 1899 by taking a sharp underwater picture at about 165 feet using a camera operated by remote control. This depth record would not be exceeded for 40 years. Several years earlier, in 1892, Boutan had said, ". . . the [marine] naturalist finds himself in a position analogous to a visitor from the moon who might make observations from a moon-ship floating on top of our atmosphere." His success in viewing sea creatures in their own realm and in bringing photographic documentation back to

" 'I love you,' I whispered
into the ear of the ocean.
'Ever since I've known you
I've loved you.
I must see all your marvels,
know all your beauty. . . .'
And the ocean listened,
and snuggled still closer to me."

Hans Hass,
Diving to Adventure

Silhouette of man darkens the sea as spearfisherman Jean Tapu slips from his outrigger canoe into a warm lagoon in Polynesia's Tuamotu Archipelago. Equipped only with a spear, Tapu dives in the tradition of his ancestors. Dropping to depths of 100 feet on a single breath, he lures fish with special sounds learned from his father. Centuries ago, in pursuit of food, man plunged beneath the waves, where a rainbow world of exotic life forms and fantastic vistas awaited him.

AL GIDDINGS / SEA FILMS, INC.

the surface changed all that. The development of undersea science has been thoroughly interwoven with that of photography ever since.

Another scientist, Dr. William H. Longley, working with National Geographic staff member Charles Martin at the Marine Laboratory of the Carnegie Institution in Florida, took the first underwater photographs in color. Using Boutan's basic approach, Longley constructed a heavy brass camera housing with external focus, speed, and shutter controls. The device could be immersed in the sea and operated by a diver wearing a Dunn helmet—a heavy, bucketlike copper apparatus supplied with air pumped from the surface. In 1926 the Longley-Martin team succeeded in taking underwater autochrome pictures; magnesium powder exploding in a pontoon raft above provided enough light for them to photograph moving fish in the sea.

COURTESY RUTH DUGAN

Underwater luminary Louis Boutan poses 12 feet down for a historic 1898 self-portrait. A zoologist and the father of undersea photography, Boutan devised a submersible camera and captured marine life on film. In the sea's dark reaches he pioneered the use of battery-powered arc lamps. About his work he wrote, ". . . I have opened the way . . . it remains for others to follow me, to open up new paths, and to arrive at ultimate success."

Undersea motion pictures, in black and white, were the creation of J. E. (Jack) Williamson, who began his career as a newspaper cartoonist. Early in the 20th century Williamson's father had built a marine salvage machine, a wide, flexible metal tube with glass viewing ports and mechanical grabs. Young Jack's still photographs—taken through the ports of the tube—were so successful that he was inspired to attempt underwater filming. His first effort, begun in 1914, was a six-reel production, *The Williamson Submarine Expedition.* He later went on to make other films, including his magnum opus, the first movie version of Jules Verne's *Twenty Thousand Leagues Under the Sea.*

For his films, Williamson created monsters of the deep from rubber and steel. Meanwhile, one of his contemporaries, naturalist Dr. William Beebe, thrilled scientist and layman alike with his firsthand descriptions of genuine sea creatures, some monsterlike in appearance, but all fascinating in Beebe's opinion.

In 1928 Beebe took thousands of readers vicariously into the clear waters off Haiti to encounter the living things there: ". . . I saw . . . an absolutely new thing to me. . . . It was floating in mid-water, oval in shape, surrounded by a band of waving fin. . . . I . . . crept after this phantom. . . . I gained on it, and saw two large eyes. . . . With the magic of sea things, he was floating gently in mid-space . . . and scarleting and palliding as nothing can in the world but a squid. . . . All his arms stretched out for an instant in my direction, and out of the transparency of the waving fin there slowly grew two more eyes, eyes more terrible than his real staring orbs. . . . After they had looked at me for a space of time which might have been a second or a week, they gradually faded away,—no whit unlike Alice's cat, and the tail fin of the squid showed again, white and unmarked."

Beebe's love for the sea and the creatures there, his passion for sharing his experiences, are expressed in his first book about diving, *Beneath Tropic Seas:* "All I ask of each reader is this,—Don't die without having borrowed, stolen, purchased or made a helmet of sorts, to glimpse for yourself this new world. Books, aquaria and glass-bottomed boats are, to such an experience, only what a time-table is to an actual tour, or what a dried, dusty bit of coral in the what-not of the best parlor is to this unsuspected realm of gorgeous life and color existing with us today on the self-same planet Earth."

Like William Beebe, zoologist Dr. Hans Hass was lured to the sea by his curiosity about the creatures that lived there—and, to some extent, by a taste for fresh fish. Like Boutan and Longley, Hass devised his own underwater camera systems—in his case for both still and motion pictures. His adventures, in turn, have enticed countless others to don masks and fins and jump into unknown waters.

It all began for him, Hass wrote, on the coast at Cap d'Antibes, France: ". . . on a sunny day in July 1937. I was eighteen. . . . As I was clambering over a large boulder I caught sight of a man swimming among the rocks, his face under water, as if searching for something in the greenish-blue depths.

"What was he diving for? As a passionate swimmer and diver I was consumed with curiosity. Was he looking for shells or sponges? Curiously I clambered on to a rock which projected far out into the sea and from there watched the strange diver at close range.

"Almost within arm's reach the sun-tanned body swam past. Then it dived, and I was able to observe it closely. . . . His long spear poised in his right hand, he swam past, below my vantage point, just like a large light-coloured predatory fish."

Hass had encountered the lithe form of Guy Gilpatric, an American writer who drew thousands into sport diving with his many reports and the classic book *The Compleat Goggler*—and who first introduced the young Hans Hass to the use of goggles and spear.

Today Hass has written numerous books and scientific papers, pioneered underwater filming, led dozens of expeditions, and founded, in Liechtenstein, the Institute for Submarine Research. Yet he smiles when he talks about that day when he was 18, eyes wide with curiosity, first watching the "man-fish" in the sea. "No doubt everybody experiences a moment in his life when he feels that he has met his own good fortune and must grasp it and hold on to it," Hass wrote. "That is how I felt that day."

In 1939 the members of what would prove to be an intrepid trio of undersea pioneers met in a small coastal village in France. Their considerable diving skills, combined with an interest in photography, led the team of Jacques-Yves Cousteau, Philippe Tailliez, and Frédéric Dumas to develop new methods and equipment that contributed to the growing field of underwater filming.

Tailliez, in his 1954 book *To Hidden Depths,* recounted: "Hunters of submarine pictures today, with all kinds of highly developed cameras available to them, perhaps have little idea of the difficulties, the disappointments, the horrible waterlogging of our apparatus which accompanied our first pioneer experiments."

In 1979 I read Tailliez's comments to Al Giddings, himself a designer and manufacturer of housings for underwater camera systems, and watched him smile wryly as he recalled quite recent "difficulties . . . disappointments . . . and horrible waterloggings."

"What is possible today, compared with 20, or even 10, years ago is staggering," Al observed. "New films and lenses have made a tremendous difference, but perhaps the most important advances have been in underwater lighting. My hero, everybody's hero, in this regard, has to be Doc Edgerton."

Dr. Harold Edgerton, beloved engineer from the Massachusetts Institute of Technology and inventor of a strobe light system, appears repeatedly in the intertwined histories of underwater scientific exploration and of photography. Edgerton inventions have been sent into the sea on cables, on submersibles, and in the hands of numerous fine photographers. His deep-sea automatic camera, designed to synchronize with a flashing strobe light and take more than 500 pictures at a charge, earned him the nickname "Papa Flash" from the crew of Cousteau's research vessel *Calypso.*

COURTESY DR. WILLIAM BEEBE

"Wonderer under sea," naturalist William Beebe wears a copper helmet to explore the reef-laced waters off Bermuda; above, he returns from a dive. In the 1930s Beebe captivated the public with tales of his adventures in the ocean depths. "As I peered down I realized I was looking toward a world of life almost as unknown as that of Mars. . ." he wrote. "In this Kingdom most of the plants are animals, the fish are friends, colors are unearthly in their shift and delicacy." To those eager to discover the "magic of water and its life," Beebe urged, "get a helmet and make all the shallows of the world your own."

"One of my favorite impressions of Edgerton dates from 1970 during one of the annual meetings of the Boston Sea Rovers," Al recalled. "Hundreds of people were assembled, as usual, after a day crammed with underwater films and talks. Some of the 'old warriors,' including Edgerton, gathered for a party, and amid the hilarity of sea stories and toasts, Edgerton, wearing a furry hat, kept standing up and taking pictures. I couldn't believe what I was seeing: Doc Edgerton, inventor of a high-speed electronic flash, was using a little camera with flashcubes! It struck me as being particularly funny that a great scientist should use such simple equipment."

Like Edgerton, photographic pioneer Dimitri Rebikoff made an important breakthrough in underwater picture-taking with a 1949 flash-tube lighting system that replaced difficult-to-handle flashbulbs. Another of his inventions, a torpedolike device for undersea photography, can be operated by a diver or by remote control.

Eye on the colorful marvels of the deep: Atop its watertight brass housing rests the camera used by ichthyologist William H. Longley to create the first underwater color photographs, called autochromes. In 1926 Longley collaborated with Charles Martin of the National Geographic Society to photograph fishes and corals off Florida's Dry Tortugas. To illuminate the seafloor, the men exploded one pound of magnesium powder for a single picture; Martin built a raft to carry the powder and designed a device to synchronize the blinding flash with an open camera shutter. "It is a gamble," Longley wrote before his experiment, and in January 1927, NATIONAL GEOGRAPHIC *published his landmark photographs of delicately hued marine life.*

SMITHSONIAN INSTITUTION
BY NATIONAL GEOGRAPHIC PHOTOGRAPHER
JOSEPH H. BAILEY

SPECIAL INDIRECT LIGHTING was required for some unusual pictures that Al Giddings took of transparent, nearly invisible planktonic creatures that had attracted zoologist Dr. William Hamner and more than a dozen students and colleagues to the blue waters off the Bahamas. Hamner, like Milne-Edwards more than a century before, wanted to meet the animals of the sea in their own realm. Previously the gelatinous creatures, with their iridescent bands and delicate trailing threads, had been caught in nets—a process that crushed the fragile plants and animals and left them in a state that more nearly resembled jelly soup than jellyfish. Hamner developed an ingenious way for scientists to observe and catch these tiny creatures by drifting with them in midwater. His divers carefully gathered plankton in jars so as not to damage the specimens, which could then be scrutinized in the laboratory. This collection process, along with the newly devised use of strobes and harmless dyes to capture the elusive organisms on film, furthered the study of plankton.

The same 1974 issue of NATIONAL GEOGRAPHIC that reported on the spectacular findings of Skylab, in a silent orbit 270 miles high, also carried Hamner's incredible observations, made in a part of the universe almost as unknown as Skylab's path, 80 feet underwater in the Gulf Stream, several miles offshore from Bimini.

"Thirty feet below," Hamner wrote, "my eye caught the outline of an extraordinary shape. . . . A large, membranous sheet spread out below me, like a transparent veil floating on a silent pool. Beneath the shimmering membrane hung a beautiful pelagic snail that resembled a large butterfly, suspended motionless in the sea." It was the first time that this creature, *Gleba cordata,* had ever been seen swimming free and the first time anyone had witnessed its web-spinning habits.

"Once thought rare and unimportant, *Gleba* is in fact an abundant and significant member of the marine food chain, because its large web traps so many tiny organisms," Hamner went on. "We wondered that plankton—so beautiful, so essential to all life in the sea, so abundant, and so fascinating in their behavior—should be so poorly known! Never before had these animals and plants been systematically observed while alive and undisturbed in their natural environment, the open ocean."

As of 1980 there are, worldwide, hundreds of scientists, photographers, and others who—as divers—seek to enjoy the ocean and the company of the countless creatures that live aquatic lives. And there are thousands more who are attracted to the sea by other incentives—

a need for food, or sometimes simply the pleasure and excitement of the chase.

Al Giddings observes, "Most of the divers I know who began more than 10 or 15 years ago were spearfishermen early on: Hans Hass, Jacques Cousteau . . ." He paused, smiled, and then continued, ". . . Al Giddings. I'll always remember as some of the most exciting times of my life the days spent swimming out more than a mile, carrying my spear and towing a raft, then diving down 50 or 60 feet into murky, cold water. I'd know that I had reached bottom when I could see the eyes of dozens of enormous lingcod resting on the sand, looking up at me as if I had just dropped in from another universe."

In 1979 Al dived in the Tuamotu Archipelago of the South Pacific, where he renewed his acquaintance with old friends who still practice the art of spearfishing.

"The inner lagoon at Rangiroa is incredibly clear, deep, and full of fish," Al recalls. "The local spearfishermen sometimes dive more than 100 feet, holding their breath for two minutes or longer. A man will swim down to about 40 feet, then sink rapidly as his lungs compress. From above he looks hauntingly like a spider artfully stalking its prey. On the bottom he'll settle down quietly, then begin to make eerie guttural trills, squeaks, and pops with his mouth and throat. The fish reacts to the sound, glides over to investigate, then—zing! It's on the end of a spear."

Al recently visited Jean Tapu, a man born in the Tuamotus, and considered one of the best spearfishermen in Polynesia. They talked about Tapu's earlier diving days, when he had been encouraged by some friends to enter an international spearfishing competition. Al told me, "The other entrants laughed at Tapu's wooden spear gun and odd-looking swim gear. They couldn't believe that he was serious about diving with such simple equipment. But when he came in with more than twice as many fish as the next highest contender, they began to believe, all right."

Tapu showed Al a project that has begun to revolutionize the gathering of one of the greatest natural treasures in the tropical seas—pearls. In a protected lagoon Tapu has established an underwater farm for cultivating the lustrous pearls characteristic of Polynesia. With Tapu, Al glided 50 feet underwater, along row after row of stakes, each strung with lines of the large, disk-shaped pearl-yielding oysters.

"We bring live oysters in from other areas and attach them here," Tapu said. "In them we implant a bead of shell and a small piece of mantle tissue and let them be for a few years. Many more pearls are formed this way than in the wild. When I started, I knew about pearls cultivated in Japan, but I only *thought* the method would work. I wasn't *sure* I could do it here."

Later, in a small hut that Tapu constructed in shallow water for cleaning and sorting the pearls, Al watched as the precious spheres were piled according to size, shape, color, and luster. Holding a perfectly round, steel-blue orb to the soft light of a lantern, he said, "It seems alive—like the living thing it came from. Tapu, I like the way your garden grows."

At present no farming methods are being applied to other natural "gems of the sea"—the precious red, pink, black, and gold corals that for centuries have been transformed by craftsmen and artists into highly polished jewelry and objets d'art.

Dr. Rick Grigg, a University of Hawaii zoologist, has spent years developing proper ways to manage precious corals to ensure their

COURTESY HANS HASS

HANS HASS

Joyous adventurer Hans Hass (top) carries the first camera he used to "hunt" undersea and reveal an undiscovered world to an eager public. In the 1930s Hass, an Austrian zoologist, began leading diving expeditions to study the creatures of the deep and to record their behavior on film. In 1939 he photographed team member Jörg Böhler catching a Caribbean turtle barehanded. A pioneer in underwater research, Hass has written numerous books on the majesty of ocean life—enticing divers to follow the slap of his flippers and enjoy the marine world.

sustained natural growth. Grigg remembers his first dive to look for the feathery "bushes" of black coral: "The cobalt blue water was featureless, empty; it looked absolutely bottomless. . . . a hundred feet down and still no bottom in sight. . . . Deeper; still deeper. At last a vast gray vista emerged from the gloom. . . . I recall a forest of black branches reaching up from the underside of a cliff face at 200 feet. Great, lacy branches, six to eight feet long, swayed gracefully in the current. . . . I drifted along, amazed, transfixed, and totally isolated from the rest of the world. I had found the subject for my master's thesis in marine ecology, and a direction for my life."

In Hawaii alone the coral industry—producing black, pink, and gold varieties—accounts for about 10 million dollars in retail sales each year and about 800 jobs, according to Grigg. Worldwide, the precious coral industry is worth an estimated 500 million dollars. But in many places the eagerness to obtain as much as possible as quickly as possible has resulted in serious depletion. To overcome this, Hawaiian companies are trying to develop sound conservation practices, largely based on Grigg's studies and recommendations.

The sea produces another kind of natural treasure—the soft cushions of tissue built by colonies of living sponges and prized for uses as diverse as washing babies and staining leather. As a child growing up on Florida's Gulf Coast, I was entranced by the boats that landed at Tarpon Springs, loaded with long strings of golden-brown puffs—partially dried "wool" sponges hooked from the clear waters offshore. Beginning in 1905, hundreds of families migrated from Greece to Florida, where untouched—and very rich—sponge beds had been discovered. For centuries men of the Aegean's Dodecanese islands had dived for sponges, using only a weight to speed their descent. By the time the sponge divers made their way to America, their methods had become increasingly sophisticated. They brought diving suits and an age-old knowledge of boats and marine life, as well as their customs, languages, foods, and music.

Today, many of the descendants of those immigrants are still in Tarpon Springs, but the thriving sponge exchange, the rows of sponge boats, and the aroma of thousands of curing animals, have largely become memories. Synthetic substitutes and a lack of interest in continuing sponge-diving traditions have turned what was once a multimillion-dollar industry for the people of the city into a tourist-based economy. The heavy bronze helmets and lead-weighted shoes that once dressed divers are now more frequently found on display as curiosities from not-too-ancient times.

SPEARFISHING, PEARL DIVING, and gathering corals and sponges were all among the earliest lures to underwater realms. But in the northwestern Pacific, on a few islands and in coastal communities of Japan, yet another diving tradition persists, little changed for two thousand years. In 1971 Al Giddings listened to a muted symphony of soft, plaintive whistles blending with the sounds of seabirds, as white-swathed forms emerged from a calm, cold ocean. Several *ama,* women divers of Japan, breathed deeply, then exhaled with the shrill but pleasant sounds that Japanese poets call *iso nageki,* the elegy of the sea.

For nearly two hours the women had made repeated excursions to the seafloor while their husbands or brothers tended their lines and looked after the abalone and snails, ribbons of brown kelp and sprigs of branching red seaweed that the women gathered from 60 feet or more below. Then it was time to return to shore, to warm up, eat lunch, and prepare for an afternoon of more diving.

Al, who photographed the ama of the island community at Hekura Jima and lived among them for several weeks, often mused about the traditional way that the women swim long hours in the sea while the men of the household attend them from boats. Whatever the origins of this unusual system, it works. The diving women either swim from shore, towing a float and a basket to hold their catch, or—like those Al dived with—work in family teams out of small boats. Traditionally, techniques are passed from mother to daughter. Even grandmothers dive, often well into their seventies.

But in Hekura Jima times are changing. Many younger women are choosing to work in factories on the mainland rather than pursue the hard life of the ama. In addition, shellfish are not as abundant as they were a few decades ago, partly because of the large amount taken. In fact, local authorities have prohibited the use of breathing aids to prevent further depletion. Today the meticulous methods of the ama must compete with the sophisticated modern technology of fishing fleets. Nevertheless, tradition lives on; the urgent need for food supports an occupation that would seem to have endured beyond its time.

As world population continues to grow and customary food sources fall short of the demand, people in some parts of the world are turning to aquaculture—the cultivation of the natural produce of the water. On land, agriculture long ago replaced hunting and gathering as a major source of food. In the ocean, looking to wild populations of fish, squid, and krill as reliable resources may be as unrealistic as trying to sustain the world's present population with game, wild roots, and berries. Already fish once thought to be "infinite" are dwindling. Dr. Sidney Holt, one of the most respected and knowledgeable fisheries biologists in the world, said in 1979: "Efforts to achieve a condition of sustainable use of living marine resources have, by and large, been unsuccessful—so far. Nature has given us some sharp warnings in recent years." He cited as examples collapses of the herring stocks in the North Sea, the recent precipitous decline of capelin stocks in the northwest Atlantic, and similar patterns for anchoveta off the coasts of Chile and Peru.

One Nova Scotia firm, Janel Fisheries, Ltd., has begun working with nature to produce fish for human consumption. Owner Jay Ettman buys bluefin tuna from local mackerel fishermen who inadvertently catch the 800-pound fish in their nets. The tuna are fattened in ocean corrals like cattle in feedlots and sold on the Japanese market for *sashimi*—a raw fish delicacy. For each pound of that tuna, thousands of pounds of marine life have been consumed in a long food chain. Knowing that it may take tons of plants to produce a tuna sandwich, I view each bite with the utmost respect! And I understand why aquaculture is increasingly important.

In Japan the paper-thin seaweed called *nori,* an alga that looks almost black when dried, is cultivated on thousands of acres of sea farms. The nori that wraps rice cakes and crackers or appears in soups was once taken primarily from intertidal rocks. Today much of it owes its existence to ocean farmers.

Cultivation of kelp, the seaweed that grows more than 100 feet long, has for years been a concern of Dr. Wheeler North of the California Institute of Technology. Colloidal substances in kelp and certain other seaweeds are used in the preparation of commodities as diverse as paint and cosmetics, dental molds and toothpaste. Because

COURTESY DR. HAROLD E. EDGERTON

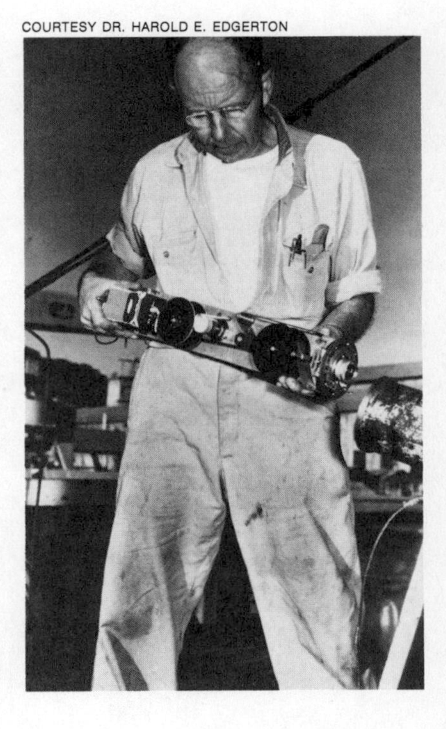

To see the unseen, engineer-inventor Dr. Harold E. Edgerton slips his remarkable deep-sea camera into a watertight housing before sending it down thousands of feet for a rare glimpse of the ocean floor. In the early 1930s Edgerton developed a high-speed electronic flash that has revolutionized photography. His achievements enabled cameras to freeze motion as swift as that of a hummingbird's wings and earned him the affectionate nickname "Papa Flash." Edgerton devices, adapted for undersea research, have plumbed marine trenches, helped to locate ancient shipwrecks, and opened new frontiers in oceanography.

of their rapid growth—as much as two feet a day—the slippery brown "sequoias of the sea" are being used experimentally as raw material for chemical conversion to methane gas. Al Giddings and I visited Dr. North to see a pilot sea farm managed by the General Electric Company off the coast of California. Dr. North helped develop the project to explore a possible alternative energy source.

We tied off our small boat near the bright orange-and-white buoy marking the site of the floating farm several miles offshore from the California Institute of Technology's marine station. At the surface I saw what appeared to be a rather ordinary-looking kelp bed, tangles of giant blades with gulls bobbing among them. But the scene underwater was unlike anything I had ever before encountered. Nutrient-rich water was being pumped from about 1,500 feet below and dispersed over the floating forest. I shivered in the flow of cold ocean water and glided with Al toward two divers adjusting the lines that held the transplanted giant kelp to a frame. We descended through a jungle of golden brown, past what appeared to be the spokes of an enormous inverted umbrella strung with the lines that anchored the plants.

I tried to imagine an open-ocean kelp farm 335 miles on each side. Scientific estimates indicate that such a farm could supply a substitute fuel to fill the natural gas requirements of the United States for a year.

World of transparent life reveals its secrets in the indigo waters off the Bahamas as marine researcher Larry Madin and colleagues (opposite) collect samples of plankton—tiny organisms that drift with ocean currents. Floating with the creatures in the open sea, a team of researchers led by zoologist Dr. William M. Hamner in 1971 became the first to use scuba in observing plankton in its own habitat. FOLLOWING PAGES: *A gossamer snail floats before the diving mask of researcher Alice Alldredge—who appears amid an array of oceangoing plankton. Harmless dyes squirted into the water enable photographers to capture the ethereal beauty of the fanciful drifters. Diving scientists study their role in the marine food chain.*

AL GIDDINGS

WHETHER OR NOT kelp farms ever come to be a major source of fuel, food, or industrial materials, the natural sea forests will, if protected, continue to sustain a complex system of life as they have for millions of years. Each square mile of California kelp has a potential worth of about a million dollars a year, Dr. North estimates, including use directly by industry and as a source of fish and shellfish. In 1834 Charles Darwin wrote, ". . . if, in any country a forest was destroyed, I do not believe nearly so many species of animals would perish as would . . . from the destruction of the kelp." Sadly, much of the kelp bordering southern California's rocky coast has been devastated by pollution in the last few decades, just as we have begun to understand the significance of these ancient ecosystems.

Since 1975 Al Giddings and I have followed the seasonal changes in the underwater metropolis associated with some pollution-free patches of kelp near Santa Catalina island, off the coast of southern California. Diversity and abundance of life are so great among the kelp fronds that one plant can harbor living creatures in numbers exceeding the entire human population of nearby Los Angeles County, some seven million individuals. A single frond may be populated by more than 100,000 small animals.

Thirty feet underwater Al and I cruised slowly, looking back at the energy-giving sunlight shafting through a canopy of kelp blades. More than 750 kinds of fishes and invertebrates, many of them micro-organisms, live among these luxuriant forests, and we saw some of the larger "citizens" after only a few minutes: graceful sea lions, golden-eyed horn sharks, fast-moving squid, and slow-gliding sea hares. I found myself face to face with an octopus parked serenely on a ledge and was glad I could linger and watch as his dark eyes met mine.

For some, the lure of the sea has tangible rewards—pearls, food, corals. But encounters such as mine with the octopus yield a different kind of treasure: new insights into the nature of creatures of the sea and into our part in the fabric of life that includes kelp and squid, crystalline jellyfish and octopuses that return our curious glances.

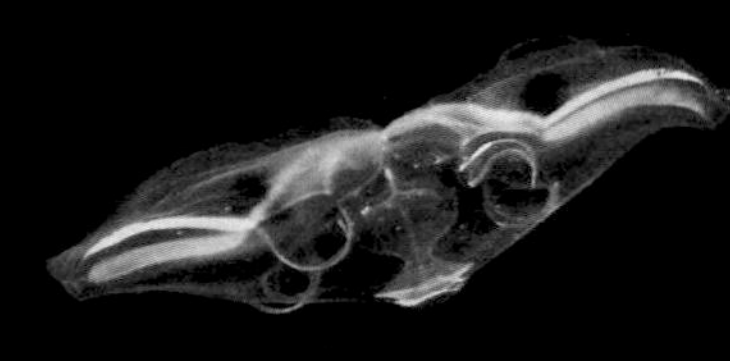

OCYROPSIS MACULATA—ABOUT 4 INCHES

NEIL SWANBERG

PELAGIC SQUID
OF THE GENUS LEACHIA—ABOUT 2 INCHES

ROGER T. HANLON / RAYMOND F. HIXON

HASTIGERINA PELAGICA—
ABOUT 3/4 INCH ACROSS SPINES

AL GIDDINGS / SEA FILMS, INC.

ALICE ALLDREDGE
OBSERVING COROLLA SPECTABILIS

AL GIDDINGS

AGALMA OKENI—ABOUT 3 INCHES

AL GIDDINGS / SEA FILMS, INC.

ORCHISTOMA PILEUS—ABOUT 2 1/2 INCHES IN DIAMETER
AL GIDDINGS / SEA FILMS, INC.

COROLLA SPECTABILIS—ABOUT 4 INCHES ACROSS WINGS
AL GIDDINGS

PEGEA CONFOEDERATA—ABOUT 2 1/2 INCHES
AL GIDDINGS / SEA FILMS, INC.

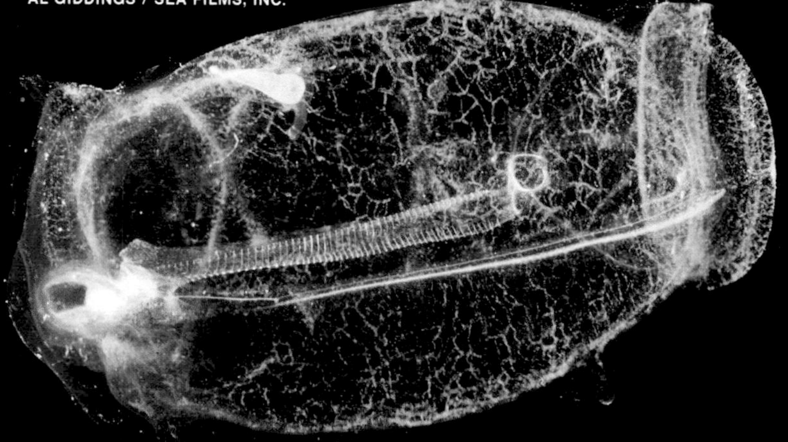

HYDROMEDUSA OF THE GENUS POLYORCHIS—ABOUT 3/4 INCH IN DIAMETER
AL GIDDINGS / SEA FILMS, INC.

Carrying a rock to speed his descent, Jean Tapu plunges through aquamarine seas of the South Pacific (below, left) in search of oysters. In the tradition of Polynesian pearl divers, he collects wild oysters in a woven rope basket (below) strung from his outrigger canoe. Men have treasured pearls for thousands of years, prizing them for their luster and beauty. Inspired by the success of the Japanese cultured-pearl industry, some Polynesians now seek ways to cultivate pearls in oysters

CHUCK NICKLIN / SEA FILMS, INC.

AL GIDDINGS / SEA FILMS, INC. (BELOW AND RIGHT)

gathered from the wild. Using a modern breathing apparatus, Tapu (right) picks oysters from his underwater farm in the Tuamotus. After implanting a bead of shell in the mollusks, he strings them from a grid and allows them to mature. Within about three years Tapu harvests the oysters and removes the shining pearls.

Sunset explodes above dark Pacific waters in the Tuamotu Archipelago. Inside his shallow-water hut, Jean Tapu sorts cultured pearls from his undersea farm, explaining their qualities to a visitor (right). Shape, hue, size, and luster—and, of course, the buyer's preference—determine a pearl's value. In a still-life arrangement on his table (far right), Tapu's prying tools cover tally sheets listing oysters by form and color. On his hat brim rest Tapu's wooden goggles, reminder of a day when he dived only for wild oysters.

AL GIDDINGS / SEA FILMS, INC.

Taking the plunge, a diver leaps into waters of the Gulf of Mexico in search of sponges (right). Garbed in a waterproof suit (above), he uses techniques perfected by Mediterranean sponge fishermen who immigrated to Tarpon Springs, Florida, after the turn of the century to harvest the Gulf's vast sponge wealth.

Encumbered by a heavy bronze helmet, the diver (far left) treads slowly along the bottom, loosening sponges with a pronged hook. Hoses connected to the boat Agatha *supply him with air. For several decades Tarpon Springs thrived as one of the world's largest sponge markets, but blights, lack of manpower, and the introduction of synthetic sponges have combined to wither the industry. At left, a worker selects sponges in the warehouse of one of the chief buyers. Of the hundreds of sponge boats that once bobbed along the Tarpon Springs waterfront, only about half a dozen remain; some sponge fishermen demonstrate the original skills for tourists, while others use modern diving equipment, updating a tradition.*

Feathery jewels of the sea, Hawaii's rare corals lure divers to twilight depths in search of the precious branches. At left, diver Harold Bloomfield hovers by an eight-foot coral tree harvested off Maui. The day's work done, a crew zips back to shore (above) with a bushy load of black coral tied astern. Most black coral thrives at depths of 100 to 350 feet, some within the grasp of skilled scuba divers. Prized above all corals, pink "Hawaiian angelskin" (right)—gathered by submersible off Molokai—lives at depths of 1,150 to 1,500 feet and grows a mere one-third of an inch a year. The scarcity of these corals concerns marine ecologists like Dr. Rick W. Grigg (far left). In an undersea garden off Maui, he records the growth of black coral and works to develop ways of ensuring the survival of these natural treasures.

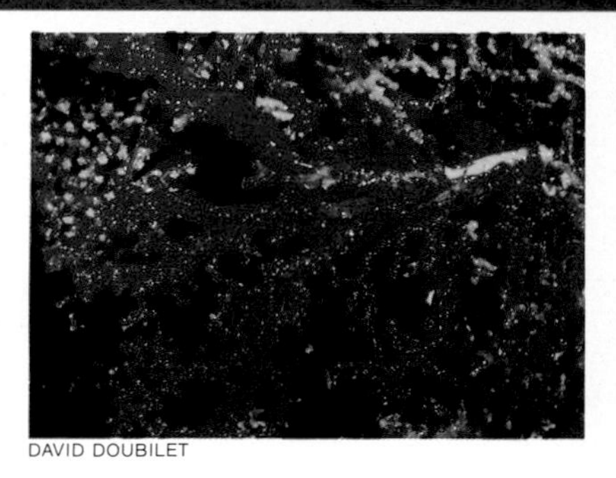

DAVID DOUBILET

Like fairy-tale mermaids, Japanese ama *fisherwomen (left) prepare for a morning of diving off Honshu's rocky coast. Without the aid of swim fins or breathing equipment, these women dive for shellfish and edible seaweeds, continuing a tradition that may date back 2,000 years. In an 1833 print featuring the remarkable ama (below), a diver surfaces with an abalone, a meaty single-shell mollusk enjoyed throughout Japan. About a hundred years ago, when once-abundant shellfish supplies began to dwindle, the ama moved to deeper waters, using goggles for better vision in the darker reaches. Diving mask on her head, a young ama (right) looks to the sea; although fewer women now choose the life of the ama, more than 7,000 divers along Japan's shores still pursue the ancient vocation.*

PRINT BY KUNIYOSHI, COURTESY B. W. ROBINSON

LUIS MARDEN

Armada of fishing boats and ghostly ama divers cluster along Japan's Pacific shore near Kuzaki, an ama village. Poised on the rocks and preparing to leap, these white-clad kachido, *or "walking people," fish only in the shallows. A lifeline hitched to her waist, an ama (left) of the* funado, *"ship people," dives from a boat anchored in deeper water. Probing into a tangle of kelp (right), she uses a sharp iron bar to pry mollusks from rocks; white cream protects her face from salt and sun. Some fisherwomen make 100 dives a day, gathering many pounds of tasty shellfish. Abalones brought from the deep (far left) hang from a bamboo rod.*

LUIS MARDEN

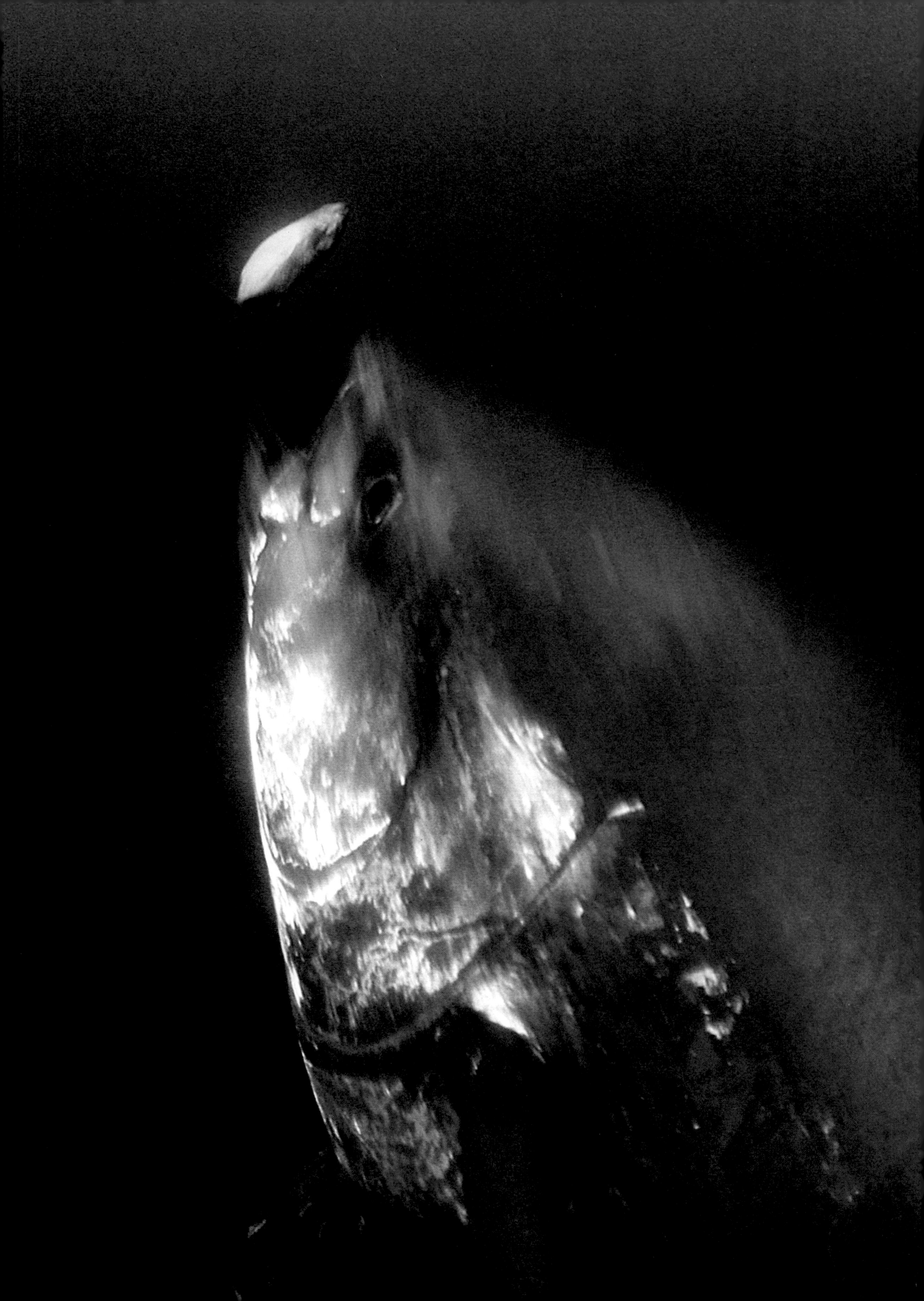

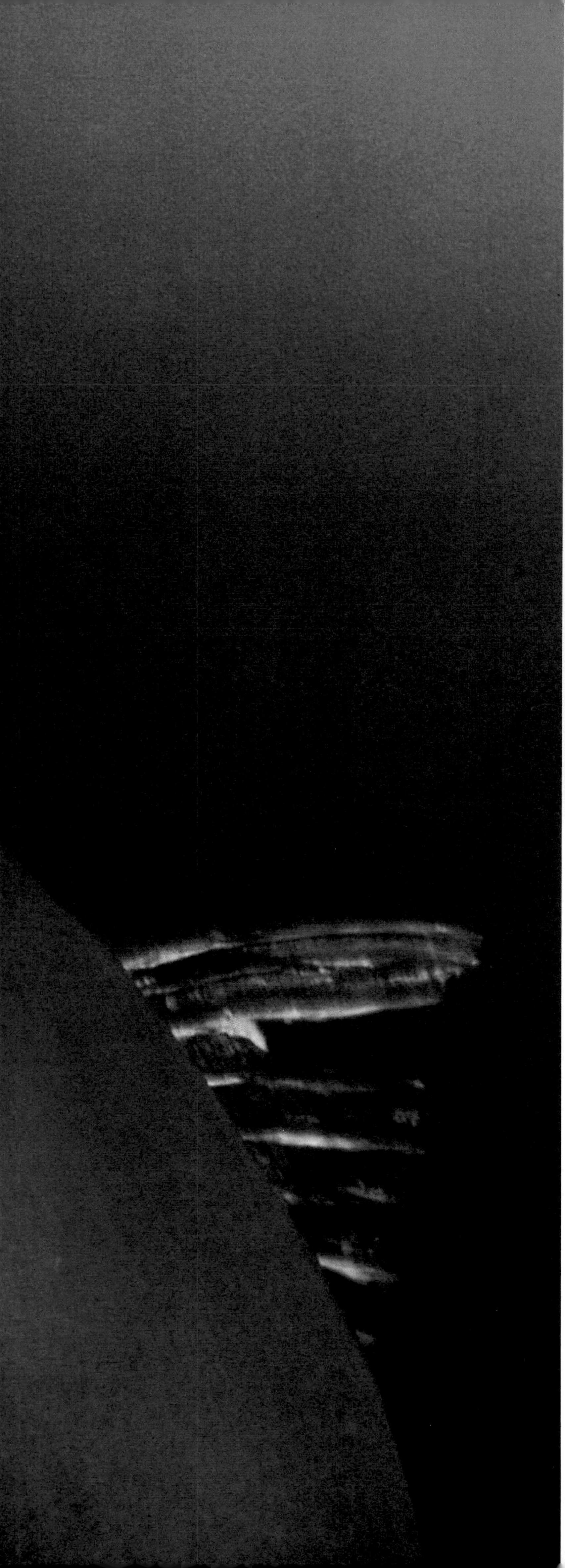

DAVID DOUBILET

Inside a fishy stockyard, a bluefin tuna (left) rockets upward to swallow a snack. Schools of huge bluefin (above)—some weighing more than 1,000 pounds—roam an underwater holding pen off the coast of Nova Scotia, where workers for Janel Fisheries, Ltd., fatten them for export to Japan. In a unique experiment with aquaculture, Janel owner Jay Ettman buys live tuna from local mackerel fishermen, whose nets trap the bluefin by accident. From June to October each captive tuna gulps down 50 pounds of mackerel and herring a day (top), growing plump for harvest in the fall. High mercury content of bluefin prevents their sale in North America, but Japanese consumers consider the fish a delicacy and pay a high price. Future marine farms may raise fish from hatchlings to adults, managing the sea's resources to feed an expanding world population.

DAVID DOUBILET

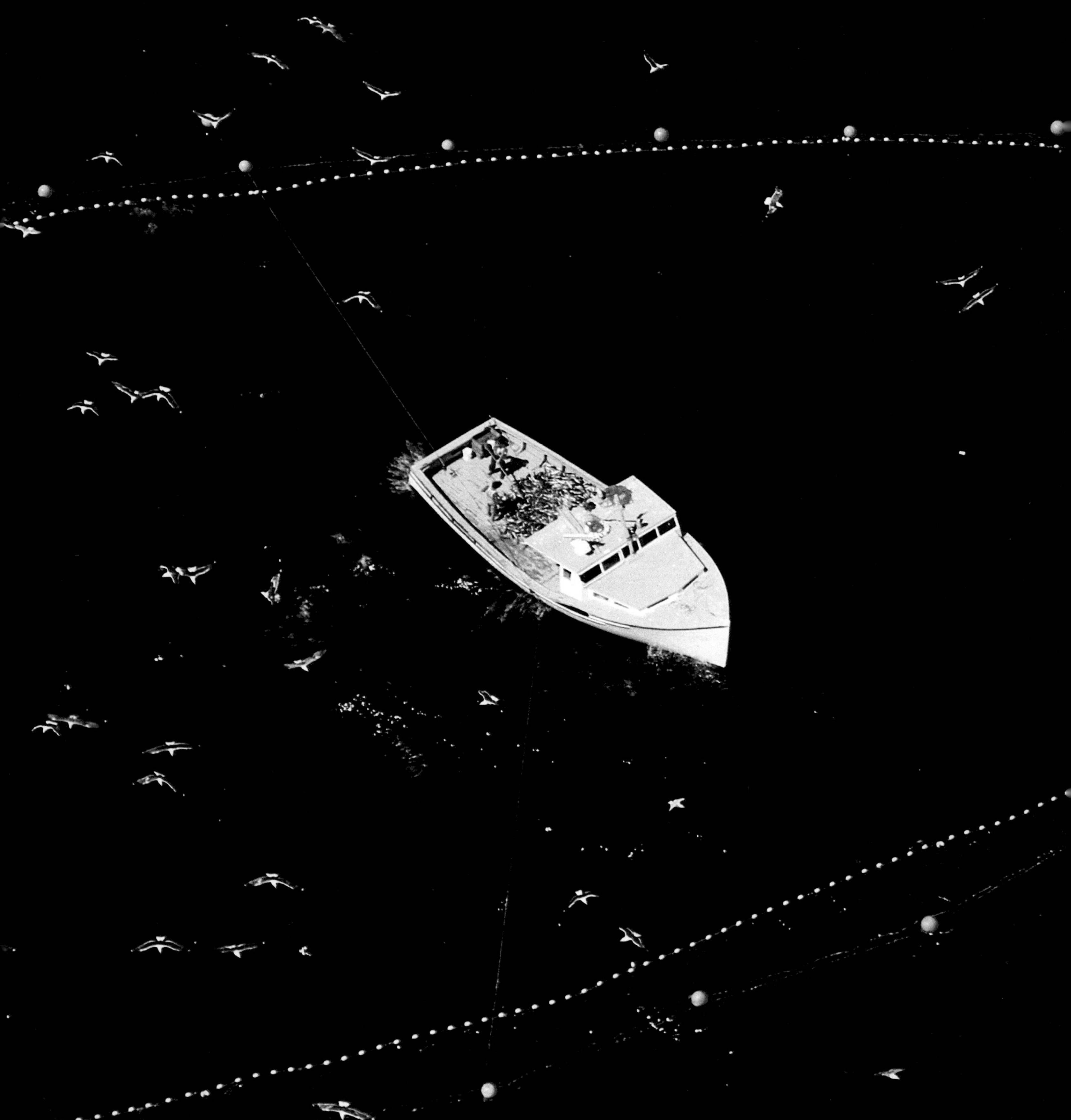

Open-water corrals, net pens marked by buoys, confine dozens of bluefin tuna as gulls circle a feed boat heaped with fish. At fall harvest Janel Fisheries flies the bluefin to Japan, where buyers at a wholesale market bid for the tuna (far left). A chef slices the marbled flesh, serving the raw fish minutes later as sashimi.

AL GIDDINGS / SEA FILMS, INC. (INCLUDING PRECEDING PAGES)

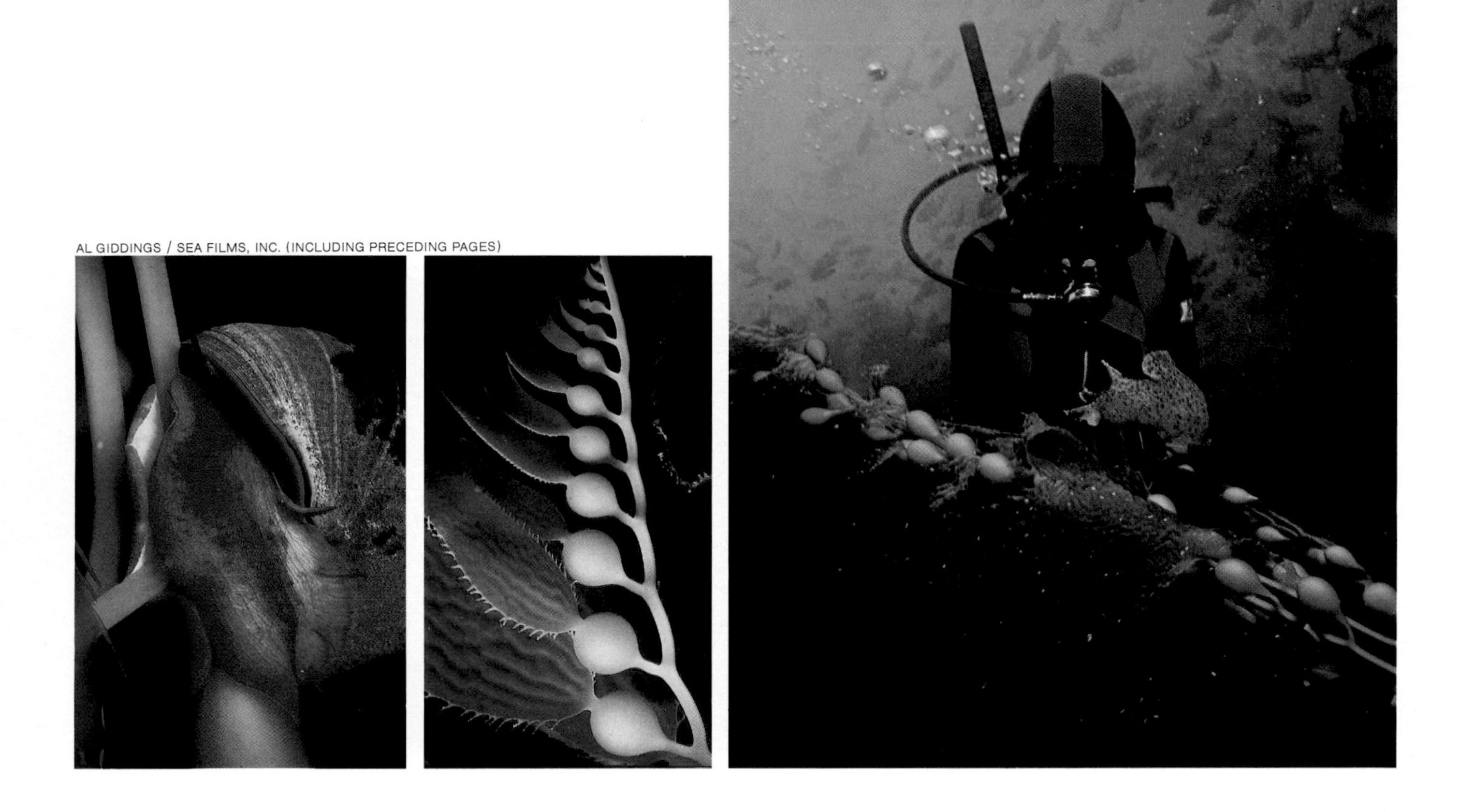

Kelp's flashiest resident, a brilliant amber garibaldi, darts among swaying fronds that grow as fast as two feet a day—reaching lengths of more than 100 feet. This submarine jungle of giant kelp (below) streams with life in coastal waters off California's Santa Catalina island. Gas-filled bladders (above, center) buoy the leggy seaweed; a tiny turban snail (above, left) grazes along a blade. A blizzard of fish swirls behind author Sylvia Earle (above, right), as she examines a speckled sea hare.

PRECEDING PAGES: *Plumes of sargassum and kelp surround the author on a research dive. Says Dr. Earle, "This is a marine botanist's dream—swimming in your favorite subject!"*

War and Pieces of Eight

IN TIMES OF WAR the sea is an indifferent stage where deadly dramas are played in earnest, in three dimensions. Fish are forgotten, beauty ignored, curiosity put aside when survival is at stake. The ocean becomes an adversary with inherent obstacles and uncertainties, but it is also a friend, providing shelter for strategic advances and invisible, equally strategic retreats.

The awesome consequences of war at sea were much on my mind one sunny day in July 1975 as Al Giddings and I descended through 250 feet of cool, blue water in Truk Lagoon in the Caroline Islands for a brief view of the remains of a Japanese warship. She lay upright on the seafloor, ghostlike, one of dozens of ships sunk during a two-day holocaust in February 1944, when 400 tons of Allied bombs and aerial torpedoes rained upon the anchored Japanese fleet. Our Trukese friend and diving companion, Kimiuo Aisek, then 17, had been terrified by the thunderous explosions.

"When the attack came," he said, "I hid in a cave on the side of Dublon Island. Several ships sank as I watched. For more than two years afterward, oil from them covered the beaches." He paused, looked at the calm water, and added, "But the sea is healed now."

I thought of Kimiuo's words as we swam to a small tank, now embroidered with encrusting corals, pink and green algae, and delicate anemones. As a botanist, I was irresistibly drawn to the extraordinary research sites that sunken ships provide as "undersea islands," time capsules that make it possible to determine something about the growth rates and development of complex tropical systems. But as a human being, I kept imagining what it must have been like in the quiet lagoon as sky and sea suddenly exploded. The heavily armed warships, now resplendent with rich growth, have undergone a sea change, mute testimony to nature's indifference to human conflict.

Nearby Al had discovered the *Shinohara,* a submarine that sank during a bombing raid in April 1944. The submarine had crash-dived with an improperly closed ventilator, and it quickly flooded.

For three decades *Shinohara* was a lost tomb. The war had ended long ago; new friendships prospered among old enemies. Al decided to try to find the sub during a summer visit to Truk in 1971. "It took days," Al told me. "Five of us went with Kimiuo and some of his friends who thought they knew about where she had gone down. Finally a suspicious image turned up on the sonar screen. One of the divers dropped overboard and came up shouting. I swam down and saw her resting upright in 150 feet of water, with her bridge blown away."

The following year Al was part of a team of American and Japanese divers who recovered the remains of the lost sailors in a setting that was itself symbolic of the futility of war.

"Methought I saw a thousand
fearful wrecks;
Ten thousand men that fishes
gnawed upon;
Wedges of gold, great anchors,
heaps of pearl,
Inestimable stones,
unvalued jewels,
All scattered in the bottom
of the sea."

William Shakespeare,
Richard III

Ballooning air bags help divers wrest a heavily encrusted cannon from the remains of Witte Leeuw—*White Lion. The Dutch East Indiaman sank in 1613 in the harbor of St. Helena, an isolated island in the South Atlantic Ocean. Homeward bound with a cargo of spices, Ming porcelain, and diamonds, the ship lost a battle with Portuguese galleons. It lay hidden beneath 110 feet of water until 1976, when undersea historian Robert Sténuit—spurred by the dual lure of* Witte Leeuw's *historical worth and its valuable treasure—mounted a successful search.*

NATIONAL GEOGRAPHIC PHOTOGRAPHER
BATES LITTLEHALES

I picked up a fragile teacup, buried in the sediment of three decades. Where would it be in the years to come? "Right here, I hope!" I thought, replacing the cup. Truk Lagoon has been declared a historical monument where visitors may look and touch, but not take.

Nearly two centuries before *Shinohara* and more than 900 other submarines were lost in the world's most devastating war, the first submarine ever to engage in combat in the United States was launched against British forces during the American Revolution. Yale graduate David Bushnell was obsessed with the desire to construct an undersea craft that could move surreptitiously among—and sink—ships menacing American waters. In 1775 he designed and built a wooden, one-man, oval-shaped submarine called *Turtle,* which boasted screw-type propulsion and a conning tower. Encouraged by George Washington, Bushnell took his machine to New York harbor. There, in September 1776, *Turtle* eased next to a British man-of-war. From inside, Sgt. Ezra Lee used a hand-powered auger to try to penetrate the hull of the large ship and attach explosives.

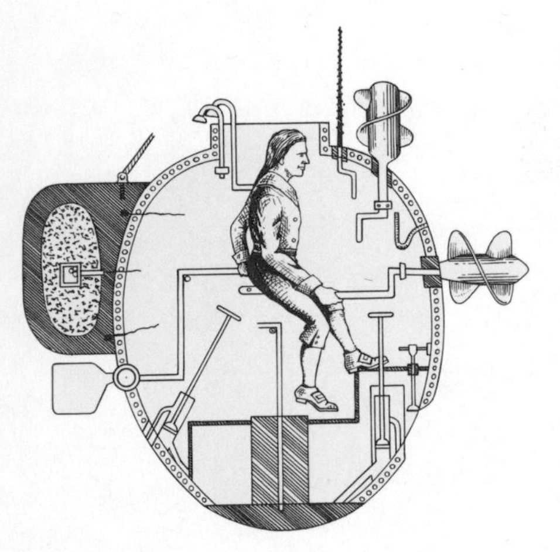

Egg-shaped submarine Turtle, *built in 1775 by Connecticut Yankee David Bushnell, fueled rebel hopes of ending Britain's naval stranglehold on New York harbor. Large enough for one man, it carried half an hour's air supply and, above its rudder, a 150-pound powder mine that the operator could detach and secure underwater to an enemy ship. Hand-turned propellers—incorrectly shown in this cutaway drawing as Archimedes' screws—drove* Turtle *forward, backward, up, or down, while bits of fox fire—phosphorescent fungi—illuminated its compass and depth gauge. The first sub used in war,* Turtle *worked well in trials but never sank a ship.*

ELECTRIC BOAT DIVISION OF GENERAL DYNAMICS

After several exhausting attempts, the combined effects of unyielding metal, faulty compass, and diminishing time brought *Turtle* and Sergeant Lee to the surface. A strong tide pushed the sub ever closer to the enemy, and *Turtle* attracted the considerable attention of several hundred men. Desperate, Lee released the 150-pound cask of gunpowder he had intended to use against the man-of-war. It surfaced, timing mechanism rattling, close to the British, who prudently decided not to investigate further. Twenty minutes after Lee safely reached shore, the powder exploded in a towering plume of water that sent the British fleet into confusion.

Artist-engineer Robert Fulton, a protégé of Benjamin Franklin, had a similar passion for designing undersea craft. Although remembered more for his steamboats than for his submarines, Fulton altruistically believed that subs were potentially so powerful that they would force the naval powers of the world to a peaceful stalemate.

In 1800, 70 years before Jules Verne's submarine *Nautilus* cut a swath through imaginary seas, Fulton's 7-by-21-foot copper-hulled *Nautilus* emerged successful after several trials. A collapsible fan-shaped sail for surface propulsion and a forward observation dome made the vessel distinctive, and other innovations included a special line to trail a torpedo. But to some detractors the idea of fighting an enemy "underwater . . . in secret" seemed not quite fair. Amid such controversy Fulton's submarine was never used in combat.

Half a century later, in 1850, Denmark's blockade of Germany prompted yet another development in underwater technology. A Bavarian, Cpl. Wilhelm Bauer, designed his submarine *Brandtaucher* (traditionally translated as Sea-Diver) in the shape of a porpoise. The Danes had never seen anything like the steel monster that approached them out of Kiel harbor. Fearing the worst, the Danish ships retreated without a shot.

Bauer launched another submarine—*Le Diable-Marin,* the Sea-Devil—in Russia in 1855, after he convinced Grand Duke Constantine that it would be useful against England in the Crimean War. A highly effective craft, *Diable-Marin* made 134 dives in depths to 150 feet over five months. The vessel also was host to the first underwater concert. A four-man brass band, with the vocal accompaniment of the sub's 13-man crew, played in honor of the coronation of Czar Alexander II—while at the bottom of Kronshtadt harbor!

Thousands of miles away, in response to the U. S. Civil War, submersibles were developed that could transport men underwater. In several battles Confederate "Davids," submarine-boat amalgams, were measured against the "Goliaths" of the North.

One of the most famous submarines was the eight-man Confederate *Hunley.* The vessel lost her crew several times during trials before she made history in 1864 as the first submarine to destroy an enemy ship in warfare. During a moonlit approach to Charleston, South Carolina, the cigar-shaped *Hunley* rammed a torpedo into the *Housatonic.* The ensuing explosion sank both ships.

Two years earlier, the Confederacy's ironclad *Merrimack* had engaged the first Union ironclad, the smaller, more maneuverable *Monitor.* The two ships sparred for hours, cannonballs bouncing off each one, with no loss of life and no appreciable damage to the ships. Finally, both withdrew.

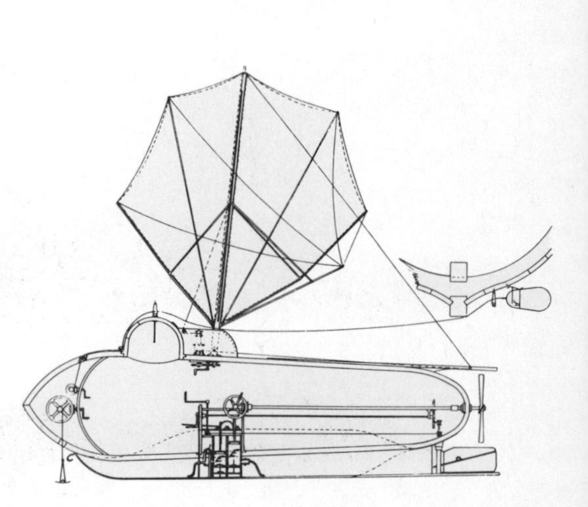

World's first Nautilus *took to the sea 70 years before Jules Verne launched his fictitious submarine of the same name in* Twenty Thousand Leagues Under the Sea. *Built by Robert Fulton, creator of the steamboat* Clermont, *this graceful, 21-foot-long design had a collapsible sail for surface travel, a forward observation dome, and room for a three-man crew. Its remarkable resemblance to modern research submersibles underscores its chief flaw: too far ahead of its time. Although Fulton proved its wartime usefulness with tests that included sinking a target ship, military powers of his day shunned* Nautilus, *considering it a coward's "threat to proper warfare."*

FROM ORIGINAL PLANS FOR ROBERT FULTON'S NAUTILUS, 1798

THE ENCOUNTER between the ironclads made a lasting impression on schoolteacher, and staunch Irish loyalist, John P. Holland. His first submarine, the *Holland I,* was built with private funds in 1878 after the Navy rejected his drawings. The sub performed so well, however, that a group of Irish expatriates, opposing the British, backed a second, more formidable weapon that came to be known as the *Fenian Ram.* Launched in 1881, the remarkable gasoline-powered submarine could move as fast as nine miles an hour on the surface and perhaps equally fast underwater.

The first seven boats of the U. S. Submarine Service all were designed and built by Holland in the early 1900s. Decades later the company he founded would build the world's first nuclear submarine, another *Nautilus.*

Concurrent, and sometimes rival, submarines were the products of another American, Simon Lake. As a child, Lake had been inspired by the adventures of Captain Nemo in Jules Verne's *Twenty Thousand Leagues Under the Sea,* and he dreamed of the day he too would travel in a submarine. The vision stayed with him. When the U. S. Navy advertised for designs, Lake submitted plans for his underwater craft *Argonaut.* The more experienced John Holland won the contract, but Lake persisted, scaling down his original design to the 14-foot *Argonaut Junior*—one of the first submarines built for peaceful purposes.

Years later Lake built the *Protector,* again drawing on private money and for the first time incorporating a practical periscope. Eventually *Protector* was sold to Russia for use in the Russo-Japanese War; other Lake subs were supplied to Austria and Germany. But not until 1912 did Lake's own country, the United States, buy one of his vessels. The *Seal* was the result. At 161 feet long the largest submarine then known, it was armed with six torpedoes and carried 24 men.

By 1917 Germany was using subs to sink both military and commercial ships. And by the time World War II was under way, the great military forces of the world were all heavily committed to using methods once regarded as not quite fair.

Today awesome fleets of nuclear-powered submarines, radically different from their predecessors, glide through the world's oceans bearing flags of several nations. Nuclear submarines have been described as the greatest deterrent to attack that man has yet achieved—a weapon that can destroy an enemy no matter how devastating his first attack might be.

These peace-by-force rationalizations, hauntingly similar to those voiced by Robert Fulton more than a century and a half ago, have spurred the development of space-age undersea craft. Now submarines are cylindrical, self-contained communities powered by a grapefruit-size chunk of uranium. Traveling submerged for months, they move from continent to continent, even under the polar ice.

The same military motives that stimulated some to build subma-

SUBMARINE FORCE LIBRARY AND MUSEUM, GROTON, CT.

UNITED PRESS INTERNATIONAL

Irish schoolteacher-turned-inventor, John Holland emerges from the hatch of his 53-foot namesake (above). Powered by a gasoline engine on the surface, Holland VI *relied on storage batteries underwater. In 1900 the U. S. Navy commissioned Holland to build six other subs, including* Grampus *(top).*

rines fired others to develop free-diving techniques. One of the first recorded instances took no special equipment, but it must have called for an unusual kind of gumption. In the fifth century B.C., a father-and-daughter team, Scyllias and Cyana, dived into the stormy Aegean Sea to loose the anchors of invading Persian ships. They raised havoc and became heroes by successfully challenging the overwhelming odds of armed enemies and a raging sea.

IN THE ADRIATIC SEA, not too far from waters where Persian fleets had sailed ages before, Italian divers Raffaele Paolucci and Raffaele Rossetti confronted the Austrian fleet in 1918. Sitting astride a miniature submarine that packed an enormous charge of explosives, the two human torpedoes succeeded in passing several Austrian barriers. Delayed by a strong tide running against them, Paolucci and Rossetti finally attached their charge to a huge battleship, *Viribus Unitis,* as dawn was breaking. Without the cover of darkness, they were discovered and, to their horror and chagrin, brought aboard the very ship they had just mined. Worse yet, the ship had been captured only hours before by their allies! The divers' warnings only partially convinced their friendly captors that the ship was doomed. Though several hundred died when the explosives went off, Paolucci and Rossetti returned to Italy as heroes.

Twenty years later, as the world's major military powers prepared for another war, the Italians again took the lead in dangerous but deadly undersea warfare by free-swimming divers. By 1942 England too had begun an extensive program of underwater tactics, using midget submarines similar to those favored by the Italian divers. Not until 1943 did the United States begin to compete with them, putting highly trained underwater demolition teams (UDT's) in the water. Some of the first UDT divers lacked even flippers and scuba tanks; they went into action with only swim goggles and a knife.

The full spectrum of underwater equipment available to military divers today would dazzle Scyllias and Cyana and would probably raise considerable comment from those who first went to sea as UDT frogmen. Silent, bubble-free rebreather systems permit hours of undersea cruising; propulsion devices speed divers to their destination with little effort; new gas mixes make it possible to dive to a thousand feet and beyond using a selection of helmets, hoods, masks, bells, chambers, and submarines with diving compartments. Scientists have also turned their attention to marine mammals and other sea creatures. At the Naval Ocean Systems Center at San Diego, California, staff members have studied the dolphin's ability to "see" underwater with bursts of clicking sounds.

In 1956 Royal Swedish Navy divers put their military underwater skills into peaceful practice when they helped retrieve a long-retired member of the Swedish fleet. *Vasa,* a 1,400-ton galleon laden with 64 new bronze cannons, had set forth on her maiden voyage from Stockholm harbor in 1628. Suddenly an unexpected breeze caught the sails, careening the majestic ship. Instead of righting herself, she listed farther, then toppled, spilling into the sea sailors, wives, and even children who were aboard to celebrate the launching.

For more than 300 years only the approximate location of the 180-foot warship was known. Then engineer Anders Franzén, while testing the harbor bottom, discovered what he thought was the *Vasa.*

Encouraged by Franzén, the Royal Swedish Navy moved its Divers Training School to the site and let students explore the wreck. Their first conversation was not promising:

"I'm standing in porridge up to my chest," a diver reported. "Can't see a thing. . . ." Several moments passed.

"Wait a minute!" he called over the wire. "I just reached out and touched something solid . . . it feels like a wall of wood! It's a big ship all right! . . . here are some square openings . . . must be gunports."

Vasa had been found.

Helmeted divers dug six tunnels beneath *Vasa*'s hull and passed steel lifting cables through them in one of the most extraordinary marine archaeological recoveries ever. Gradually the ship was eased into a shallower place, where she lay for two years while work continued underwater. After making some repairs, divers attached inflatable pontoons to the keel so the ship could be floated to the surface.

I saw *Vasa* several years later, not on the bottom of Stockholm harbor, but in a remarkable, specially built shelter along the shore. Retrieved intact and brought back to the realm of light and air, *Vasa* looked little the worse for wear after her long sleep in the sea. Low salinity in the harbor had prevented wood-borers from destroying her hull. The wooden timbers remain for the thousands who have come to see and touch a part of Sweden's history.

The history of recovering things lost at sea, whether a pocketknife or a treasure-laden ship, is probably as old as the history of man's adventures in, on, and around the ocean. Ships from mainland Greece have been carrying trade goods back and forth to the Aegean Islands for at least 9,000 years. There are no complete records on the types of craft used by early traders—whether skin floats, dugouts, rafts, or boats—or on the number swallowed by the sea. A conservative calculation of 10 ships a year would total 90,000 for one area. But in a single year, 480 B.C., more than 600 Persian ships perished in two storms. Centuries later the voyage of Columbus to America in 1492 opened trade routes to the New World. Hundreds of ships sank along these sea paths, carrying to the bottom vast hoards of treasure, including such Spanish coins as doubloons and gold and silver pieces of eight.

Perhaps the single most valuable find was on the *San Pedro,* a 16th-century wreck off Bermuda. Mendel Peterson, then Curator of

THE MARINERS MUSEUM, NEWPORT NEWS, VA.

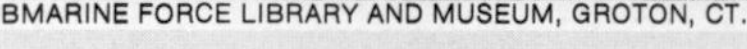

SUBMARINE FORCE LIBRARY AND MUSEUM, GROTON, CT.

John Holland's chief competitor, Simon Lake, favored wheeled submersibles that theoretically could roll effortlessly along the ocean bottom. Despite that misconception, Lake contributed to undersea research with a workable periscope and a lock-out hatch.

Naval History at the Smithsonian Institution in Washington, D. C., was among the first to view the treasured item. "The green fire dazzled me," he said. "Seven emeralds, set in a softly glowing gold cross, burned with an ethereal light. . . ." His inventory of other valuables continued: "gold ingots . . . each weighing almost two pounds . . . a

ENGRAVING FROM *HARPER'S WEEKLY* OF JAN. 24, 1863

gold bar a foot long weighing 2 1/2 pounds . . . two small sections from a similar bar which had been chiseled off to 'make change' . . . delicate gold filigree buttons set with pink conch pearls . . . a mass of silver coins, blackened from the action of the chemicals in the sea water."

Despite such spectacular finds of sunken treasure, undersea archaeology was largely ignored by the academic world until 1960, when George Bass, then a graduate student at the University of Pennsylvania, made believers out of his more conventional colleagues by successfully applying traditional archaeological techniques in the sea. The burly, dynamic Bass attacked the problem in his characteristically thorough fashion. He measured positions, took photographic records, and produced other documentation under 91 feet of water. The object: the oldest shipwreck known, a Bronze Age cargo ship discovered by archaeologist Peter Throckmorton off the coast of Turkey.

Bass, clearly a believer in the principle that "what goes down must also come up—but only after it has been thoroughly documented, *in situ,*" has now led numerous expeditions and engaged hundreds of colleagues to study dozens of ships.

The summer of 1973 proved a severe test—and confirmation—of Bass's determination. In two days he dived on five wrecks that represented four different periods of antiquity. It was a bonanza, after weeks of false leads and stormy weather had made it seem that the search would end without notable success. Two of the historically significant wrecks were lying so close together that they could be excavated simultaneously. Though side by side, they had settled to the seafloor hundreds of years apart. One, "the glass wreck," contained not only glass bowls, decanters, and plates, but also numerous raw glass ingots that flashed purple and green and yellow when held to the light.

Before he led his first underwater excavation in 1960, George Bass had not dived more than 10 feet—and that in a Pennsylvania

THE MARINERS MUSEUM, NEWPORT NEWS, VA.

Reeling in high seas off stormy Cape Hatteras, Monitor *(left) sinks on December 31, 1862, as the side-wheeler* Rhode Island *stands by helplessly. Only 10 months earlier the Union ironclad had heralded an end to the age of wooden ships by battling* Merrimack *to a historic draw. Sailors aboard (right) also served important blockade duty. Sought by salvagers for more than a century,* Monitor *eluded detection until 1973, when a Duke University research vessel, armed with sonar and underwater cameras, located the rusted, upside-down remains of the famed "cheesebox on a raft."*

swimming pool. Within a few years he had an impressive record of diving experience and an inventory of underwater paraphernalia that included a submersible decompression chamber for long dives.

SOME SEARCH FOR HISTORY, others search for treasure. Their objectives sometimes conflict. Archaeologists, as scientific detectives trying to solve the mysteries of our past, are understandably upset when treasure hunters make off with important clues that therefore cannot be studied. Yet on occasion the lure of treasure meshes with historic considerations. Robert Sténuit's interest in salvaging the Dutch East Indiaman, *Witte Leeuw*—White Lion—was twofold. As a devoted marine historian fascinated by the burgeoning trade between Europe and Asia, Sténuit was eager to reveal more about the early period of colonial expansion. One way to do so was by studying the contents of the 17th- and 18th-century East Indiamen. But through extensive research he also knew that a valuable cargo—1,311 diamonds, along with other jewelry and spices—was aboard.

Months of work in the deep waters of the South Atlantic near St. Helena Island not only confirmed the identity of *Witte Leeuw* but also revealed some unexpected cargo unlisted on the ship's manifest. Ming porcelain—fragments at first, then complete masterpieces—were found buried among tons of peppercorns.

Bowls, dishes, wine cups, pitchers, and jars surfaced—but no diamonds. Finally, after seven months, a letter arrived that made further search for jewelry and other artifacts stowed in the ship's elusive aft section futile. A South African historian had located a new eyewitness account of the sinking of *Witte Leeuw,* clearly stating that the aft part of her "blew up all to pieces." But Sténuit admits no regret. The *Witte Leeuw*'s collection of artifacts and porcelain, now on display in Amsterdam's Rijksmuseum, provides an invaluable look into times past, a priceless perspective today. Acknowledges Sténuit, "That is *Witte Leeuw*'s greatest treasure."

Similar wealth was discovered in two Spanish galleons that were wrecked in a hurricane off the coast of the island of Hispaniola. The ships were bound for the New World when they sank with more than 650 passengers and a huge and varied cargo. The *Tolosa* and *Guadalupe* have been called "supermarkets where you can find all the goods that are necessary for daily living"—if you lived in the 18th century. Carried along with the ordinary workaday things were exquisite jewelry, fine glassware, and numerous other luxury items.

JONATHAN S. BLAIR (BELOW AND OPPOSITE)

Glowing with an iridescent corrosion bestowed by the sea's thousand-year embrace, a delicate glass flask delights diver Yüksel Eğdemir, who found it in the remains of a medieval cargo ship sunk off Turkey. Expedition leader Dr. George F. Bass (opposite) glides over the amphora-littered wreck; a metal grid permits precise location of each find. An innovator, Bass applied grids and other standard archaeological techniques to undersea excavations.

Al Giddings, who witnessed the recovery, said: "It seemed that the most lavish pieces from the finest jewelers in the world had been collected in one place. One pendant had 22 diamonds and 8 emeralds; another was a simple gold brooch—with 37 diamonds."

But the most extraordinary treasure aboard the two ships was, like the sea that held it, a liquid—mercury. Numerous casks of the metal were being transported for use in processing silver and gold from New World ore. By 1976 the leather and wooden containers had nearly rotted, leaving gleaming pools in the remains of the ship's hold.

Tolosa carried 150 tons of mercury, valued today at more than five dollars a pound. Coupled with other treasures aboard, this ship rivals the greatest underwater finds in history—in material wealth.

ANOTHER SMALL SHIP, with no known gold or mercury aboard, is nonetheless also regarded as one of the great finds of the century. She is the *Monitor,* the same Civil War ironclad that inspired submarine-builder John Holland in the late 1800s.

The morning of December 31, 1862, *Monitor* sank in a violent storm off Cape Hatteras, North Carolina, a place notorious as the "Graveyard of the Atlantic." Decades of unsuccessful attempts to find the *Monitor* finally came to an end after months of research in 1973. Scientists aboard Duke University's research vessel *Eastward* used a side-scan sonar to locate the long-sought ironclad. Standing by, smiling broadly, was Dr. Harold Edgerton, Massachusetts Institute of Technology's inventor of the sonar device. Later films by an underwater television camera identified the ship.

Four years later divers breathing a mixture of helium and oxygen glided forth from the belly of a space-age submersible, the *Johnson-Sea-Link,* developed by Edwin Link for just such a mission of exploration. The juxtaposition of the submerged hull of the ironclad "cheesebox on a raft" and Link's sophisticated acrylic-and-aluminum creation symbolized the magnitude of change wrought in a century. Under 220 feet of cold, gray ocean one vessel, equipped with radio and sonar, gently touched the other, whose last desperate signal was made with a handheld lantern.

Recovering the ship for a final onshore resting place was carefully considered—and rejected for the moment. Too fragile to lift without breaking, *Monitor,* still submerged, was designated the first marine sanctuary in the United States by the Department of Commerce in 1975. It was a noble gesture for a noble ship and a hopeful sign that a spirit of caring travels with civilization's giant leaps of technology.

AL GIDDINGS

RON AND VALERIE TAYLOR (BELOW)

AL GIDDINGS / SEA FILMS, INC. (BELOW)

Life sprouts from war's relics in Truk Lagoon, where American air raids in 1944 sank some 60 Japanese vessels, creating the rusty skeleton of what now comprises the world's largest man-made reef. Corals, sponges, algae, and other growth blanket the

wrecks, turning protuberances such as the bow gun of the Japanese warship San Francisco Maru *(above) into fertile gardens and playgrounds for clouds of blue damselfish. Author Sylvia Earle (top, left) watches as bubbles of oil rise from one hulk. Gas masks and other detritus of war (far left) add to Truk's spooky display, as does a human skull (left). Unlike natural reefs, which cannot be accurately dated, these stem from a specific time, thus affording scientists valuable information on coral growth rates and colonization patterns. Congress has declared Truk Lagoon a historical monument, thereby protecting it from souvenir hunters.* FOLLOWING PAGES: *Fish swarm in the bridge section of the sunken* Fujikawa Maru.

AL GIDDINGS (FOLLOWING PAGES)

AL GIDDINGS / SEA FILMS, INC.

No ordinary swimming hole, the U. S. Navy's Submarine Escape Training Tower (left) at Pearl Harbor offers military personnel a controlled, clear-water "ocean" in which to learn basic techniques of underwater survival. Eighteen feet in diameter and 134 feet high, this aboveground tower can hold a water column 119 feet deep. Here aspiring Navy divers and submariners practice free ascents—coming up without scuba gear—from 50 or more feet deep, as they might from a disabled submarine. Each dons a Steinke hood (above), an air-filled life jacket that extends over the head to permit normal breathing while it buoys the wearer to the surface. Devised by Lt. Comdr. Harris Steinke, the hood has remained standard equipment on Navy submarines since 1964.

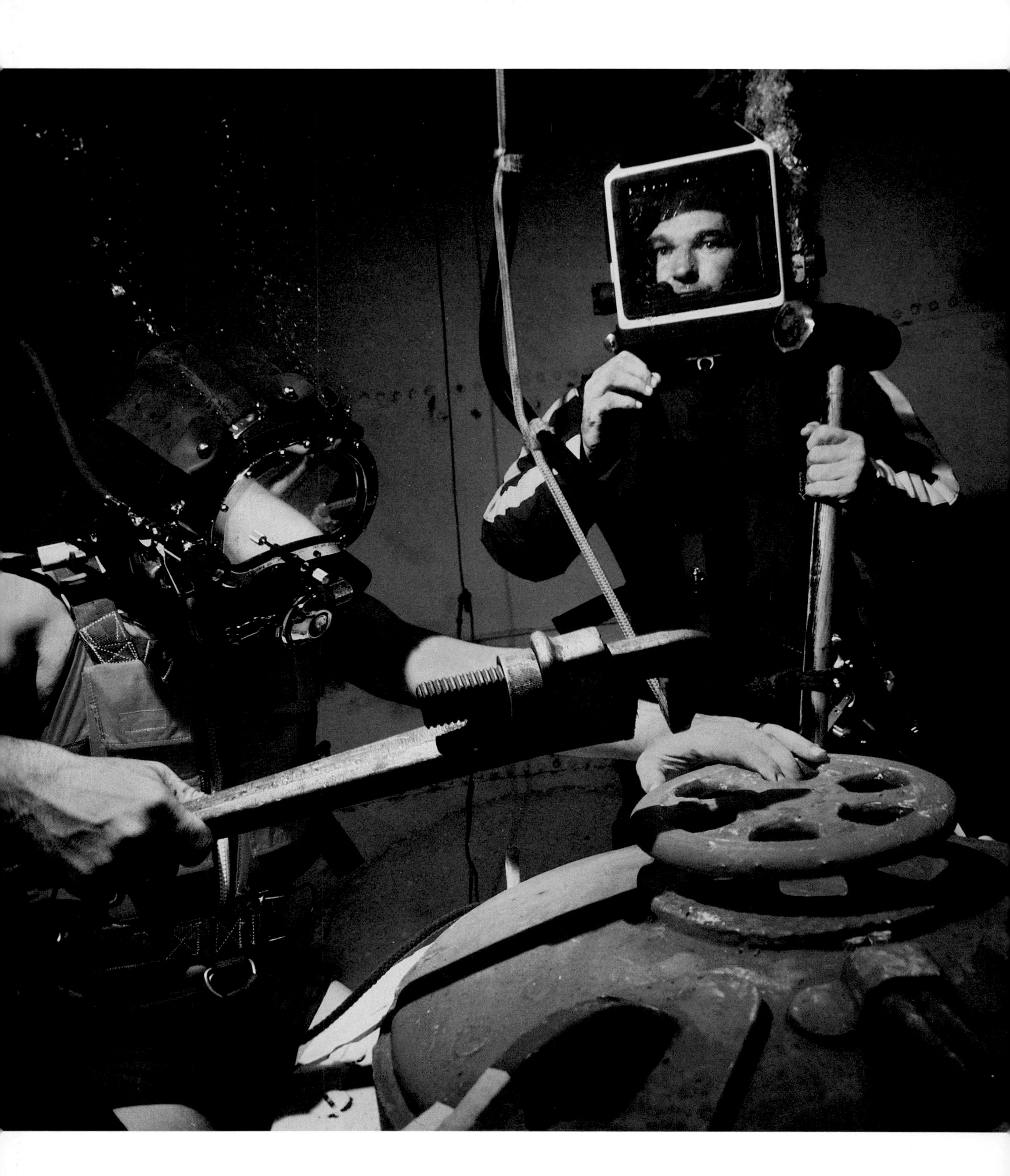

Divers to the rescue: Armed with a wrench, a sledgehammer, and modern underwater breathing gear, two Navy men descend to the bottom of the training tower and simulate a lifesaving operation—opening a submarine's jammed hatch. In addition to such survival exercises, the Navy carries out research with dolphins (FOLLOWING PAGES) in the hope of learning more about the animals' incredibly effective system of echolocation. They also explore possible military roles for the animals, which could include surveillance and detection of the enemy.

DAVID DOUBILET (FOLLOWING PAGES)

destroyed the ship's oak timbers. Divers found Vasa *in 1956; salvagers raised her in one piece by passing six-inch cables beneath her partially buried hull. Transferred from seafloor to drydock (below, right), the warship underwent months of exhaustive restoration work while sprinklers kept her long-soaked timbers from drying out and crumbling. Today she rides permanently at anchor in a Stockholm museum.*

W. E. ROSCHER

Broken timbers and modern-day refuse surround expedition leader Robert Sténuit (below) as he probes the Witte Leeuw *wreck site with an air lift—basically an oversize vacuum cleaner. "Once past an initial layer of mud," he wrote, "we came on a stratum of coral mixed with an incredible assortment of refuse—beer bottles, tin cans, old shoes, dinnerware. . . ." Of prime interest to the divers: 1,311 diamonds listed on the ship's manifest. In their stead the treasure seekers recovered a wealth of Ming porcelain (right), unmentioned on the manifest and protected in* Witte Leeuw*'s spice-packed hold. Slowly, a piece at a time, numerous fragments and dozens of intact bowls, dishes, wine cups, drinking pots, pitchers, and jars came to light. As for the gems, none ever surfaced; the explosion that shattered the vessel's stern just before it sank probably dispersed them.*

N.G.S. PHOTOGRAPHER BATES LITTLEHALES

N.G.S. PHOTOGRAPHER BATES LITTLEHALES

Precipitous stairway leads from St. Helena's cliffs to Jamestown and James Bay (above), where Witte Leeuw *exploded and sank in 1613. Details on a bronze cannon found at the site (top, left) confirm the*

wreck's identity by revealing the name of the United Dutch East India Company, to which the Indiaman belonged. Archaeologist Sténuit catalogs part of the take halfway through his seven-month expedition (far left) and carefully cleans a wine pot (left). Fortunately the wreck lay in calm water, and surface weather generally remained mild. But Sténuit and his divers had to plumb 110-foot depths repeatedly, requiring long periods of decompression. The value of the porcelains shown on these pages? From $100 to $1,000 each. Their discovery—linked obviously to the year of Witte Leeuw*'s fateful voyage—provided ceramics experts with an important starting point for analyzing and dating Ming Chinese porcelains, work long made difficult by the unvarying Ming styles and by imitation among potters.*

HARBOR BRANCH FOUNDATION, INC. (BELOW AND RIGHT)

Rising from the ocean underworld, an acrylic-and-aluminum creation called Johnson-Sea-Link I *(right) surfaces off Cape Hatteras after voyaging 220 feet to depths that conceal the remains of U.S.S.* Monitor. *In addition to surveying and photographing the famed Civil War ironclad, this four-man submersible helped divers raise such relics as a brass lantern with red lens (above, left) that some believe gave the final distress signal just before* Monitor *sank. Barnacles and other sea life encrust a section of the ship's half-inch-thick steel deck plate (above, right), transformed by time into an artificial reef on the otherwise sandy bottom. Pictures of the still-submerged turret (below, left) and a handwheel in its engine room (below, right) reveal similarly lush growth. Special equipment aboard* Johnson-Sea-Link *enabled researchers to make precise photomosaics and three-dimensional composites of the wreck, despite its treacherous surroundings, where unpredictable weather and currents often caused underwater visibility to vary from 5 to 120 feet.*

STEVE BOWERMAN / SEA FILMS, INC. (OPPOSITE)

GORDON P. WATTS, JR., (BELOW AND RIGHT)

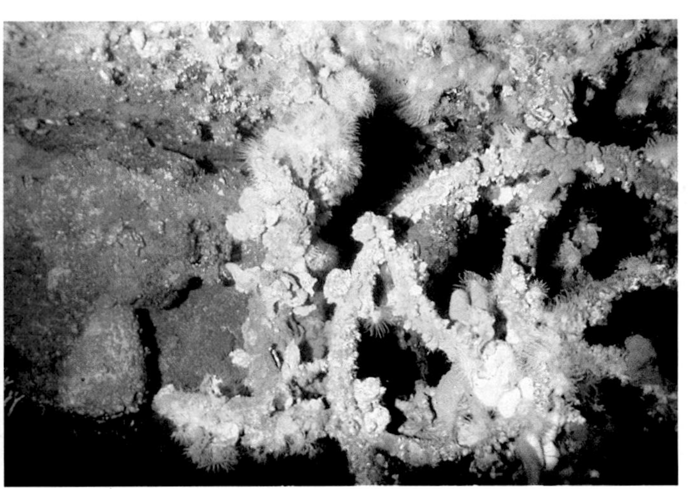

Divers employed by Caribe Salvage collect still-bright mercury from Conde de Tolosa, *a Spanish galleon that sank in 1724 off what is now the Dominican Republic. It and another galleon,* Nuestra Señora de Guadalupe, *succumbed to the same hurricane while carrying the liquid metal —used in refining silver and gold ores—to Spain's American colonies. At right, the salvagers explore* Tolosa's *ravaged skeleton, inspect clay jugs once filled with supplies, and return with their finds to the salvage ship* Hickory *(left).* PRECEDING PAGES: *A diver hovers in the glow of undersea lights during a night descent.*

JONATHAN S. BLAIR (INCLUDING PRECEDING PAGES)

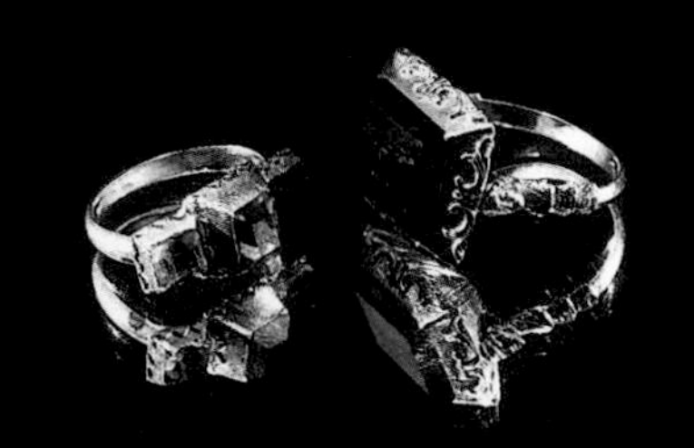

Treasures from the Tolosa *and* Guadalupe, *far more valuable than their mercury cargoes, include a varied hoard of personal wealth brought along by well-to-do passengers. Caribe Salvage divers surfaced with rings, silver coins, and doubloons (left), as well as pearls and an assortment of gold brooches and pendants (below). The diamond-encircled medal in the center bears the cross of the chivalric Order of Santiago. Dice (bottom, left) enlivened gentlemen's idle hours, while* higas, *jewelry shaped like miniature stone fists (bottom), fended off the evil eye. Also discovered: goblets, tumblers, and elegant tableware, all in perfect condition after their long sea burial.*

PEDRO BORRELL (BELOW); JONATHAN S. BLAIR

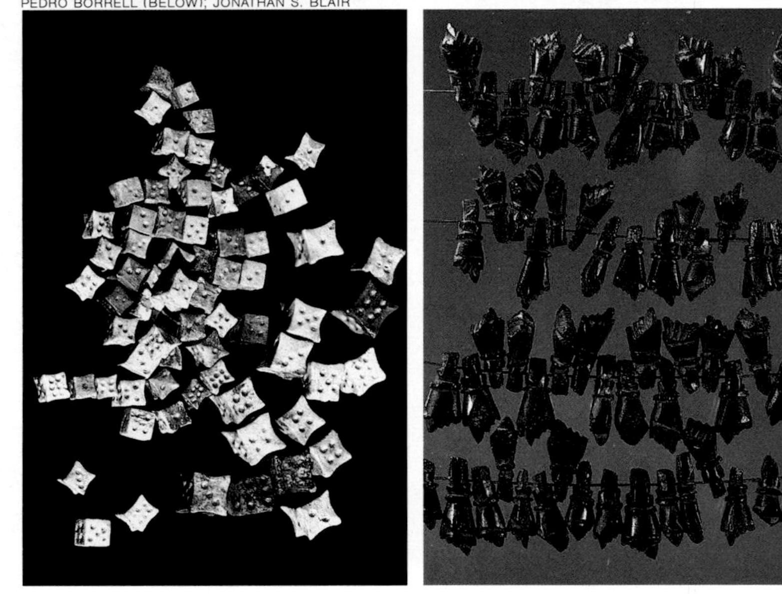

Scores of clay jars from the wrecks undergo cleaning in Casas Reales Museum (right) in Santo Domingo; some held wine or water for the long transatlantic voyage. As agreed before the search for the two galleons began, Caribe Salvage and the Dominican government equally divided the finds. Discoveries of such shipwrecks provide historians with valuable clues about ships in which man has traveled the seas as well as about life-styles of the past.

Into the Depths

Holding my breath, I gave a thrust with my flippers and dived to watch the dark form of a 40-foot-long humpback whale recede into the blue-black depths 100, 150, 200 feet below, then disappear into a realm where humans are rare visitors. Another whale approached, and for a moment I saw three air-breathing, warm-blooded mammals swimming together as my two diving partners, Al Giddings and Chuck Nicklin, joined it. Then the second whale dived deep, well beyond the range of the scuba equipment Chuck and Al were wearing.

I surfaced after less than a minute, cleared my snorkel, gulped a few breaths, and dived again to listen to the eerie sounds made by these whales during their annual sojourn in Hawaiian waters. The whales remained submerged for 11 minutes before their great black backs broke the surface a hundred yards away. A mist sprayed skyward as they exhaled; then they inhaled and dived again.

I have often wished to slip into the skin of a whale, to go where the whales go, see what they see, sense what they sense, day and night. I try to imagine what sperm whales experience when they dive for more than an hour, going several thousand feet into the eternally dark, near-freezing waters of the ocean. Being mammals like us, how are they suited to function in the depths? How do they keep warm? Find their way around? Stay in touch with one another?

Throughout time earthbound humans have admired and been inspired by hawks soaring gracefully on currents of air. Now the sight of the whales awakened in me a similar irrepressible desire to loose terrestrial bonds, go beyond nature's limits, glide on ocean currents—to be at home in the sea.

But how? Our bodies are remarkably versatile, useful for walking, running, climbing. But we are not, by nature, aquatic. Holding our breath gives us limited access to the underwater realm. Or we may take air with us in containers, like underwater astronauts, submerging ourselves within metal or glass bells, spheres, or cylinders. Only modern technology and remarkable ingenuity enable us, for a while, to meet fishes and marine mammals in their own environment, as if we too were undersea residents.

Marine mammals are adapted to overcome the problems that confound human divers—pressure, for example. On land we live at the bottom of a sea of air. At sea level the pressure of this air—"one atmosphere"—is 14.7 pounds per square inch. The thin air at the summit of Mount Everest, 29,028 feet up, exerts a pressure of about 4.5 pounds per square inch. And when balloonist Auguste Piccard ascended more than 10 miles, he experienced little more than a pound of pressure per square inch.

"How do you get to the great depths . . . ? How do you return to the surface of the ocean? And how do you maintain yourselves in the requisite medium? Am I asking too much?"
Jules Verne,
Twenty Thousand Leagues Under the Sea

From out of the gloom an awesome sight approaches: a 30-ton humpback whale. Though a mammal, like man, the humpback has a body that equips it for survival in the underwater world. The animal can dive for 20 to 30 minutes, going to depths of several hundred feet. Its thick layer of blubber helps insulate it from the numbing temperatures of the sea.

SYLVIA A. EARLE

Underwater we have to cope with the reverse situation. In calculating the effects of pressure on a diver as he descends into the sea, we start with a surface pressure of one atmosphere. For each additional 33 feet the diver descends, another atmosphere—14.7 pounds per square inch—is added. When a sperm whale dives a mile below the surface of the sea, the weight of the water exerts more than a ton of pressure on each square inch of its enormous body. Yet it is not crushed, because living tissue contains mostly water, and water, unlike air, is difficult to compress. For this reason there is no real sensation of pressure underwater, except on air spaces in the body.

For divers, problems arise mainly with lungs and sinuses—cavities filled with air—which are affected by changes in pressure. Intense ear pain halts many divers at about 10 feet. To stop the discomfort and avoid a broken eardrum, I pause and "pop" my ears by swallowing, or I hold my nose and gently blow to equalize the pressure.

A fine analogy to the way the body responds to pressure was

presented by the dean of diving historians, Sir Robert Davis, in his classic *Deep Diving and Submarine Operations:* "If such a thing as a plum or grape be lowered into deep water and pulled up again, its delicate skin will be found to have sustained no damage because the watery contents are incompressible . . . but if an empty airtight tobacco tin be

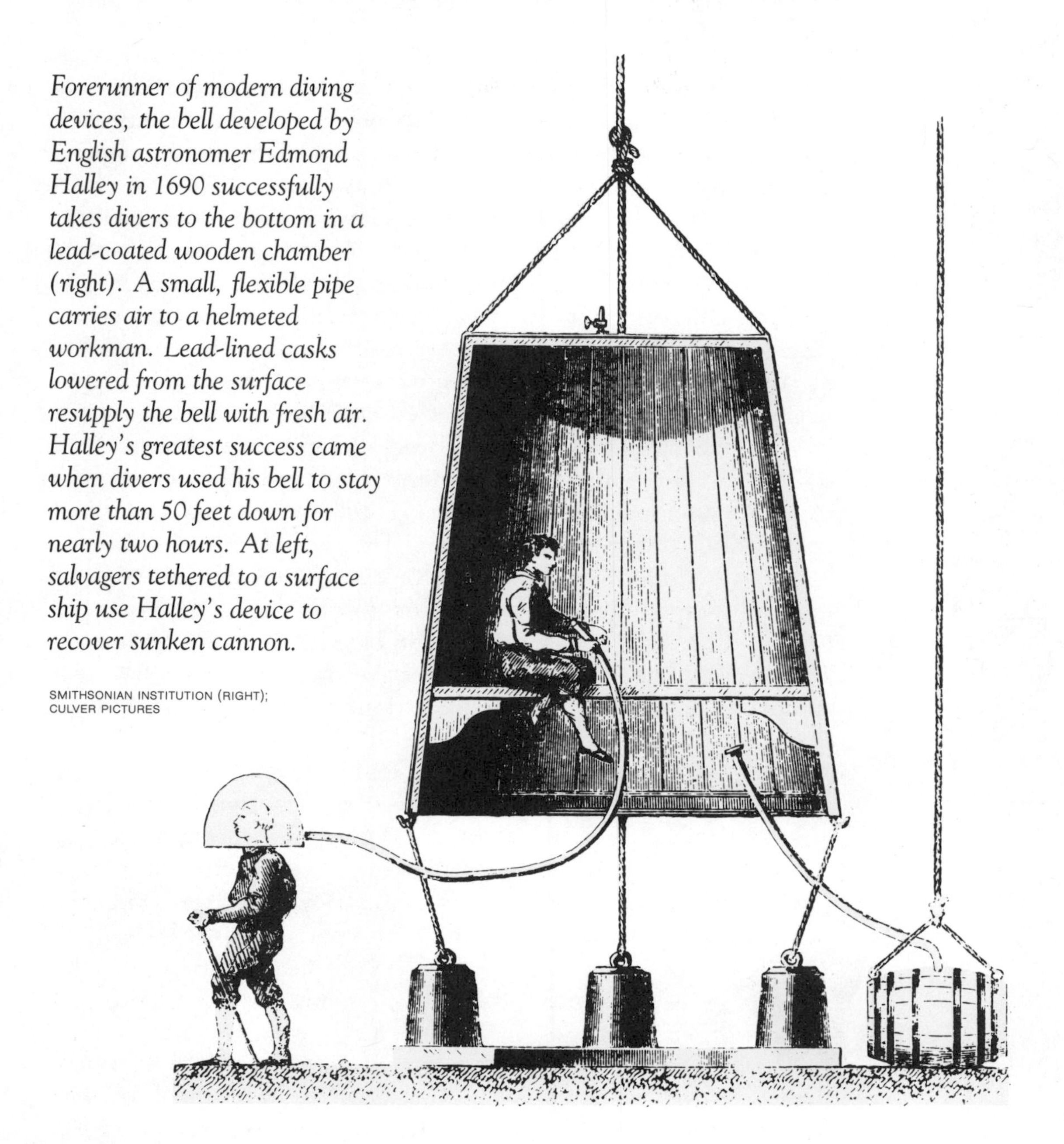

Forerunner of modern diving devices, the bell developed by English astronomer Edmond Halley in 1690 successfully takes divers to the bottom in a lead-coated wooden chamber (right). A small, flexible pipe carries air to a helmeted workman. Lead-lined casks lowered from the surface resupply the bell with fresh air. Halley's greatest success came when divers used his bell to stay more than 50 feet down for nearly two hours. At left, salvagers tethered to a surface ship use Halley's device to recover sunken cannon.

SMITHSONIAN INSTITUTION (RIGHT); CULVER PICTURES

lowered, it will be crushed by the pressure, because the air it contains is compressible and yields when the pressure is transmitted to it through the thin tin walls."

Such matters must be taken into account in building any submersible. To get to the depths and back safely, the vessels must be specially designed. The deep-diving bathysphere created by naturalist William Beebe and engineer Otis Barton underwent 1,360.3 pounds of pressure per square inch when it was suspended on a cable in 3,028 feet of water; each window endured some 19.22 tons of water. *Trieste,* the Piccard bathyscaph, sustained pressure of 16,000 pounds per square inch when it finally settled nearly seven miles underwater on the bottom of the Mariana Trench, off Guam. There, Jacques Piccard and Lt. Don Walsh, USN, looked out of the portholes to see the eyes of a flounderlike fish staring back—proof that familiar life forms can and do live even in the greatest depths of the ocean, surviving the most immense weight of water the sea has to offer.

SMITHSONIAN INSTITUTION

CULVER PICTURES

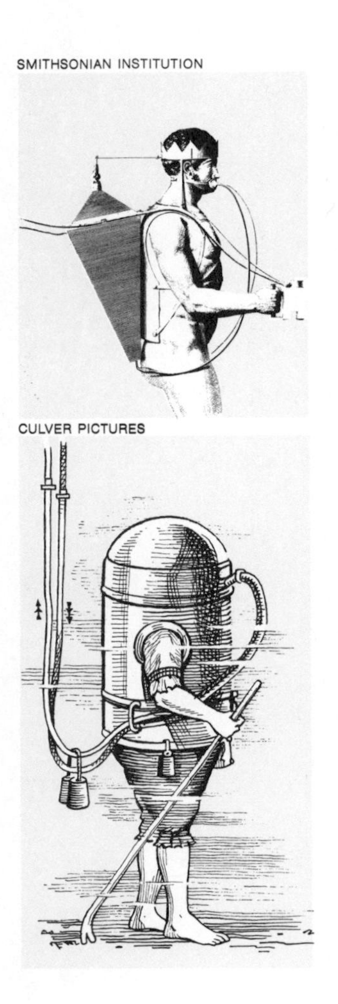

If they are to probe the ocean depths effectively, divers must overcome another of nature's underwater obstacles: numbing cold. Most of the sea is too chilly for human comfort, even at the surface. Temperatures decrease as you plunge deeper. Thousands of feet down, ocean temperatures worldwide reach a near-freezing mark. But one needn't dive deep to experience the problems caused by loss of body heat. Comparatively warm water, 86°F (30°C), can be fatal if exposure is prolonged, though we are completely comfortable and safe at such air temperatures. The reason is simply that water is an extraordinarily efficient conductor of heat. In water colder than body temperature, an unprotected human loses warmth two to five times faster than he does in air of the same temperature. How fast the person loses heat is dependent in part on how fast blood circulates to the body surface when it is in contact with the surrounding sea. Once deep-body temperature falls below 95°F (35°C), malfunctions such as temporary amnesia may occur. When body temperature is in the range of 86° to 89.6°F (30-32°C) an inactive diver may lose consciousness after hours of exposure and in some cases may even suffer heart failure.

Marine mammals and some other sea creatures are supremely well equipped to deal with low temperatures. Dolphins, whales, and even seabirds have a sophisticated internal heating system that I have often envied while shivering in a cold sea. Their arteries, like ours, carry blood away from their trunks to their extremities. A closely placed network of veins then absorbs heat from the arteries before it is lost to the sea. The veins transport the warm blood back to the body core. This heat-exchange system is so effective that dolphins and whales can overheat through prolonged exertion or lengthy exposure to air that is quite comfortable for us.

Sea otters, polar bears, fur seals, and penguins have evolved an additional system: a covering of fur or feathers that traps an insulating layer of air between skin and sea. It's a design we've copied with the diver's dry suit, which is more cumbersome, but more effective, than a standard wet suit. The latter insulates the body with a very thin layer of water trapped between suit and skin and warmed by the body.

WE'VE BORROWED OTHER IDEAS from whales and fishes. Leonardo da Vinci sketched designs for finlike hand paddles in the 16th century, and Benjamin Franklin, well known for his inventiveness as well as his statesmanship, made his own version of swim flippers. However, most divers contented themselves with bare feet until French inventor Louis de Corlieu designed, patented, and produced rubber foot fins that first appeared on the market in 1935. Since they greatly increase a swimmer's power, it is no wonder that fins—with numerous variations—have rapidly become popular with swimmers worldwide.

Breathing tubes too could have been inspired by the blowholes of dolphins and whales, or perhaps independent ingenuity was responsible for their use throughout history. In any case, a diver with a snorkel enjoys an aquatic mammal's advantage of being able to draw a breath without lifting the head out of the water.

Lost in history is the exact origin of the diving mask. The human eye sees only blurry images when exposed directly to the sea, but when it is protected in some way there is a startling difference in underwater visibility, even in a swimming pool. According to one report, Persian pearl divers in the 1300s used a mask of thin tortoiseshell. Modern divers, most commonly, wear a faceplate or goggles. Today's diving masks usually cover both the eyes and nose—a design that helps to keep the mask from flattening under pressure.

In *The Silent World* Jacques Cousteau described his reaction to donning goggles: "One Sunday morning in 1936 at Le Mourillon, near Toulon, I waded into the Mediterranean and looked into it through Fernez goggles. I was a regular Navy gunner, a good swimmer interested only in perfecting my crawl style. The sea was merely a salty obstacle that burned my eyes. I was astounded by what I saw . . . rocks covered with green, brown and silver forests of algae and fishes unknown to me, swimming in crystalline water. Standing up to breathe I saw a trolley car, people, electric-light poles. I put my eyes under again and civilization vanished with one last bow."

Diving with only primitive aids is still practiced in some parts of the world. Jean Tapu, born just a few years before Cousteau first opened his eyes to the Mediterranean, continues to test man's limits with the most ancient and widely used method of approaching the sea: breath-hold diving. Using hand-carved wooden goggles fitted around small pieces of glass, and depending on his strong will and great ingenuity, Tapu dives in the waters near his birthplace, the Tuamotu Archipelago of the South Pacific.

In late September 1979 Al Giddings sat on the edge of an outrigger canoe with Tapu, preparing to dive. For Tapu it would be one of many such excursions—for Al, an unusual opportunity to photograph the old ways of spearfishing. Outboard motors and fiberglass boats are replacing wooden canoes, and scuba gear is often used by residents and visitors alike in the Tuamotus.

But for Jean Tapu breath-hold diving is almost as easy as breathing itself. When he dives, he slips naturally into the sea. To conserve energy and speed his descent, he sometimes carries a rock for weight, a technique also used by sponge divers in the Mediterranean and by the ama divers of Japan.

At 10 feet Tapu is as deep as the deep end of many swimming pools. Most people pause here to allow air pressure in their sinuses and ear canals to equalize with outside pressure. But Tapu dives so often that his ears adjust quickly, and he continues steadily downward.

At 33 feet, with pressure double that at the surface, Tapu's smooth, brown body ripples. He sinks with increasing speed, but still exerts no effort as he descends, dropping ever deeper. On the bottom, at 75 feet, he is subjected to more than three times surface pressure. Half a minute has passed, Al notes. For most swimmers it would be uncomfortable to hold back a breath at this point. But Tapu looks unconcerned as he settles quietly near a large coral mound.

Fifty seconds.

He waits motionless. A silver-sided jack glides near the coral, and Tapu, a statue suddenly alive, darts swiftly forward, neatly taking the fish with a wooden spear gun. Then, without apparent urgency, he gracefully begins his ascent.

Seventy seconds.

Dark forms of several gray reef sharks, attracted to the struggling fish, emerge from the edge of visibility and move in rapidly. Tapu kicks for the surface, breaks through, and lifts the fish high out of the water, glancing first at Al, then below.

"Ours, not yours, this time!" he says, addressing the sharks.

Ninety seconds have elapsed since Tapu plunged in—not that unusual for Tapu, or for others who breath-hold dive frequently. Ama divers on excursions to 100 feet may stay underwater for a minute and

19TH-CENTURY ENGRAVING FROM HACHETTE

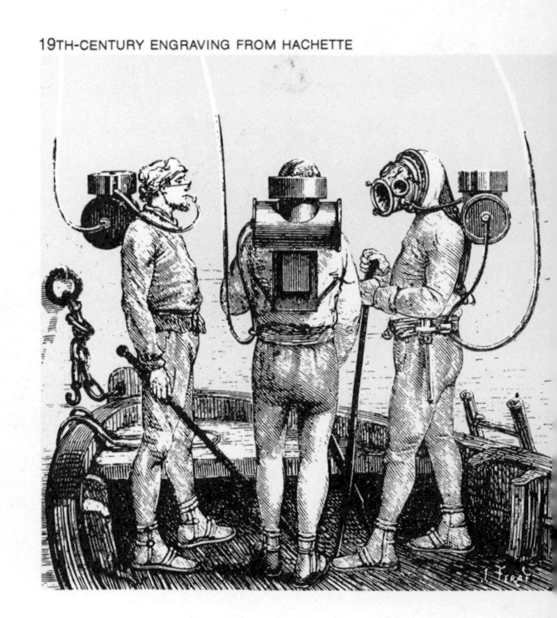

Precursor of the Aqua-Lung, invented by Benoît Rouquayrol and Auguste Denayrouze in the mid-1800s, equipped divers with canisters of air; tubes from a surface pump kept the canisters full. Captain Nemo mentions this gear in Jules Verne's novel Twenty Thousand Leagues Under the Sea. *Half a century before, Frédéric de Drieberg developed an apparatus (opposite, upper) that required a diver to nod his head continuously; rods attached to his crown pumped bellows on his back, compressing air for him to breathe. An earlier device, designed in Germany in the 1700s, featured a large helmet with armholes. Tubes carried fresh air in and foul air out.*

a half to two minutes. And in 1913 a Greek sponge diver, Stotti Georghios, swam to 200 feet to put a line on a lost anchor. During his dive he used no fins and experienced seven atmospheres of pressure—about a hundred pounds per square inch.

One explanation for such remarkable diving abilities and resis-

Late 19th-century innovation—a recompression chamber—offers relief from one of diving's deadliest hazards: the bends. At right, a diver tests a suit by entering a 1913 diving chamber that simulates pressures of the deep. Enclosed in heavy suits, divers of the period (far right) cleaned and repaired ship bottoms, did salvage work, and built piers, seawalls, and breakwaters.

DRAGER

tance to numbing temperatures at these depths is a shift of blood from the body extremities to the thorax. This phenomenon allows some breath-hold divers to go much deeper without lung damage than would be expected from measuring their lung capacity alone.

In 1968 a U. S. Navy diver, Robert Croft, performed a series of dives near Bimini in the Bahama Islands designed to measure breath-hold diving capability. He allowed doctors to make detailed medical notes and proceeded to set a record for the deepest breath-hold dive up to that time: 240 feet.

Al Giddings witnessed the dive: "I went down using scuba to 250 feet to await Croft's arrival. At that depth the nitrogen in the compressed air I was breathing had a dizzying effect on me. It was incredible enough as it was, suspended in the open sea, an ethereal infinity of blue in all directions. Then Croft came racing down the guideline, skin loose, chest compressed from the pressure. After reaching 240 feet he let go of the weight and hand over hand—no fins—pulled himself back to the surface. It was remarkable seeing him get down but really amazing to watch him go back to the surface under his own power." Croft averaged 3.95 feet per second going down and 3.5 feet per second returning.

A Frenchman, Jacques Mayol, used a somewhat different technique in 1976 to go to 328 feet and return after a total breath-holding time of 3 minutes, 40 seconds. Wearing flippers and special underwater contact lenses, Mayol used a weighted sled to descend to a marker in 105 seconds. There he activated a self-inflating buoy that lifted him back to the surface in 103 seconds.

For 15 years Mayol has studied dolphins and whales and experimented with breath-hold diving to find ways to extend his underwater capabilities. He practices yoga in preparation for his deep excursions. Before his record-setting dive he was relaxed, full of anticipation.

"When I am down there," he said, "I'm not a man anymore. I'm a sea creature . . . a diving mammal. I belong to the water."

Dives exceeding four minutes have been recorded, but oxygen deprivation much longer than this can be damaging—or fatal. In the Tuamotus, those who make successive, lengthy dives to great depths risk a condition they call *taravana*, a sickness that includes vertigo, nausea, partial or complete paralysis, and unconsciousness. But most

are careful not to abuse their bodies, and many islanders continue regular diving well beyond age 60.

A diver from the Tuamotus who at 59 went to 100 feet as many as 50 times a day summed up his attitude toward this skill in a 1962 article in National Geographic: "It is nothing. . . . I have big lungs and a strong body. It is my work."

Two minutes, three, four—a long time if you are holding your breath, but what if you are trying to follow fish? Or turn the frozen valve on an offshore oil rig? Or ease an amphora from a watery resting place? Or wrest gold coins from the embrace of ancient sediment? I try to imagine how much we would know of mountains, deserts, or forests if we had to explore these wilderness areas the way we explore the undersea world. How many of us, standing at the edge of a forest, could run in, locate, catch, and bring back to our starting place a rabbit, or a bird, or a handful of seeds—without breathing? Yet that is the difficulty facing spearfishermen and others who dive for food. The obstacles

Modern diver in historic garb trusts his luck to a diving hourglass reconstructed from 16th-century drawings. The inverted glass globe will trap a supply of air for a descent by Pete Romano into California coastal waters. Italian mathematician Niccolò Tartaglia published the design for the bell in 1551.

CHUCK NICKLIN / SEA FILMS, INC.

keeping us from the unknown forests of the sea have challenged minds through the ages and have resulted in some bizarre successes—and failures—in underwater exploration.

THE FIRST RECORDED ATTEMPTS to prolong time underwater employed primitive diving bells—inverted containers that trapped large bubbles of air. Aristotle described such inventions in the fourth century B.C.: ". . . they enable the divers to respire equally well by letting down a cauldron; for this does not fill with water, but retains the air, for it is forced down straight into the water. . . ."

According to various sources, Alexander the Great, one of Aristotle's pupils, descended into the sea in a diving chamber. Several versions of Alexander's dive have been suggested by historians and artists, but all point to a chamber, a basic step in diving technology that seems to have changed little for centuries. About 1240 English philosopher-scientist Roger Bacon reportedly described "instruments whereby men can walk on sea or river beds without danger to themselves." A few centuries later, in 1535, a diving bell designed by Italian Guglielmo de Lorena supplied air to divers for salvage excursions. The mission, successfully accomplished, was to locate sunken barges in Lake Nemi, near Rome.

Other versions of the de Lorena bell followed, all applying the same general principle, one that is easy to demonstrate. If you turn a drinking glass upside down and push it underwater, the air stays trapped inside. Similarly, a diving bell traps breathable air inside. It is safe until oxygen is depleted or the buildup of exhaled carbon dioxide becomes toxic. But the deeper the bell goes, the more pressure is exerted on it. And, just as in a diver's lungs, the air trapped in the bell compresses with increasing water pressure.

At a depth of 33 feet pressure is twice that experienced on the surface. Therefore, the air in a diving bell—or in the lungs of a breath-holding diver—will be compressed to occupy only half its accustomed space. At 99 feet—four atmospheres of pressure—volume is one-fourth that at the surface.

Throughout the 1600s diving machines of various sorts were developed, some supplied with air pumped from the surface. One of the most extraordinary-looking devices was designed and used for practical salvage operations by an Englishman, John Lethbridge, in 1715. His wooden cylinder had a glass window for viewing and armholes that permitted the person within to have his hands free for working. By the end of the 1700s many bells and other kinds of diving equipment were in use throughout Europe for salvage, construction, exploration, and even as sight-seeing attractions.

One of the 19th century's most widely used devices for underwater construction was the caisson, a huge stationary chamber with one end resting on the seafloor and the other at the surface. Divers had access to the bottom work site through a dry compartment at the top. Caissons could hold several workers at once and allow them to stay below for an extended shift—but not in safety. Long-term caisson diving resulted in maladies that ranged from mild itching and painful joints to paralysis and even death—all symptoms of caisson disease. Not until early in the 20th century did the cause of the condition—commonly called the bends—become widely known.

Diving chambers are fine—as far as they go. But they lack mobility. The need to move about freely spurred the invention of numerous devices, including several fired in the *(Continued on page 103)*

DANIEL NORD / SEA FILMS, INC. (ABOVE); CHUCK NICKLIN / SEA FILMS, INC.

Recreating early salvage work, diver Pete Romano demonstrates the use of Tartaglia's device in 30 feet of water (center tier). Pete's role: a 16th-century diver. His task: to retrieve a silver crown buried in the sand near an old anchor. From the left, Pete takes a deep breath and swims toward the treasure. Working quickly, he recovers the crown from the sand and returns to the bell. Safe inside, he examines his prize while breathing from a store of compressed air. The small reservoir allowed divers to stay down only a few minutes. In another recreation, diver Jack Monestier (left) crouches beneath a facsimile of the Sturmius diving bell invented in the 17th century. At bottom, Jack retrieves a swivel gun and hauls it

onto a platform. A diver in the Sturmius bell could stay below for as long as 20 minutes. Underwater devices have intrigued men of genius through the centuries. In the fourth century B.C. *Aristotle described how ". . . divers . . . respire equally well by letting down a cauldron; for this . . . retains the air. . . ." Roman scholar Pliny the Elder wrote about military divers who breathed through tubes, one end in their mouths, the other above the surface. Renaissance artist-inventor Leonardo da Vinci devised a snorkel for use by Venetians during a war against Turkey, but the Venetians turned down the design. They pointed out that the enemy would see the ends of the tubes sticking above the water, eliminating the element of surprise. Leonardo then designed a self-contained diving apparatus, but he suppressed the drawings: ". . . I do not publish or divulge these [plans]," he wrote, "by reason of the evil nature of man, who would use these as a means of murder at the bottom of the sea."*

Spouting clouds of bubbles, divers hover 10 feet below the surface near Grand Cayman, an island in the Caribbean. To avoid decompression sickness, divers must surface gradually, allowing nitrogen in their bodies to dissipate. The longer and deeper the dive, the slower the ascent. These divers, who have plunged to 90 feet, wait here for seven minutes. The diver at right carries an extra tank of air, in case someone runs short. Photographer Al Giddings (below) whiles away the time during a decompression stop by blowing rings of compressed air.

BERNIE CAMPOLI (LEFT); SYLVIA A. EARLE

fertile mind of Leonardo da Vinci. One of his models had a man breathing through a tube to the surface, where the tube was buoyed by a float—a long snorkel, in effect. It looked fine on paper, but in practice the external pressure on a person's lungs made it impossible to draw air from even a few feet below the surface.

In 1819 the resourceful Augustus Siebe designed his "open-dress" diving suit, which resembled a diving bell in the way it operated. The suit consisted of a helmet attached to a jacket that covered the diver to the waist; air pumped into the helmet by force escaped at the jacket bottom. After the suit was used successfully for almost 20 years, Siebe modified his original design to an airtight "closed-dress" suit—the classic hard-hat diving suit used in underwater salvage and construction. Siebe's new dress completely enclosed the diver; only the hands were left uncovered for working.

Writer-diver Robert Marx records the following 1859 account of a first-time diver's experience in the hard-hat suit: ". . . it is a large grey garment, made of india-rubber, all in one piece, and of course waterproof. One has to get into it from above, as if into a sack, and it is finished off by a pair of trousers with feet, as well as two sleeves. . . . I was then shod with a heavy pair of shoes with leaden soles, each weighing ten pounds. . . . my shoulders were loaded with the metallic helmet collar. . . ."

The diver described how two 40-pound lead weights were hung, fore and aft, before he was lowered over the side. There, he added, ". . . I felt myself beat about and buoyed up by the natural movement of the waves rolling one over the other, in spite of the leaden weights attached to me. . . . I could scarcely summon up presence of mind enough to observe the gradual deterioration of the light round me; it was a pale, doubtful twilight, which to me very much resembled the London atmosphere in a November fog. . . . At last . . . I found my feet were resting on a surface which was something like solid."

Working in such a suit is possible but confining and awkward. Little wonder that sooner or later less cumbersome systems did away with the long, vulnerable lengths of breathing hoses that supplied air to divers below.

Uneasy peace prevails between a diver and a sleek gray reef shark in waters off the Maldives in the Indian Ocean. Though sharks rank among the most feared of ocean creatures, this one has grown accustomed to handouts of fish. Despite a strongly developed sense of territory, reef sharks here tolerate human intruders, at least as long as they continue to bring food.

AL GIDDINGS / SEA FILMS, INC.

THE CONTRAST BETWEEN the experience of the hard-hat diver and the sensations described by a free-swimming diver suggests a difference as great as that of an armored knight on his steed from a bareback rider. Lotte Hass, wife of Austrian zoologist-explorer and underwater pioneer Hans Hass, described her first dive in *Girl on the Ocean Floor.* It took place in the Red Sea, where she used an early version of an oxygen rebreathing system:

"I slid down the anchor cable hand over hand. The weight of my body was balanced by lead weights so that I was just slightly heavier than the water. . . .

"I looked around me. I had already seen similar coral reefs, but that had been a swimmer's view—in other words, a bird's eye view. But now I was in the midst of this fantastic fairy garden. . . . A shoal of blue fish had engulfed me and were circling round me at close range. . . . Gradually my frantically beating heart calmed down and a wonderful feeling of joy came over me. I was in another, foreign element, the beauty and strangeness of which beggared all description."

Despite 19th-century advances in diving suits and diving bells, access to the water remained limited and dangerous as long as the bends were unexplained. A vital clue had been provided in the 17th

NATIONAL GEOGRAPHIC PHOTOGRAPHER
EMORY KRISTOF

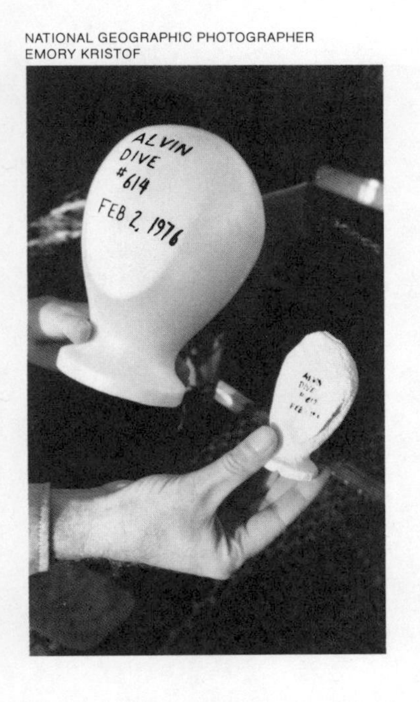

Shrunken head—once a full-size wig stand like its companion—illustrates the force of water pressure in the depths. The head collapsed after a descent to 12,000 feet in a nonpressurized compartment of the submersible Alvin. *There, pressure reached more than two and a half tons per square inch. Readjustment to lessening pressure can cause the bends—so named for the contortions of victims. At right, Larry Joline staggers after diving on an archaeological site off Turkey in 1961. He later spent 38 hours in a recompression chamber but emerged only partially recovered.*

century by scientist Robert Boyle. Boyle had long been intrigued by the effects of atmospheric changes on animals. In a series of experiments he placed animals in a pneumatic vacuum chamber and then rapidly returned them to room atmosphere. When he then examined them, he found bubbles forming in their blood and other body tissues—bubbles that he described but could not explain.

In the 1870s a French doctor, Paul Bert, connected these enigmatic bubbles with what he knew from studying the effects of pressure on alpinists at high altitudes and on divers at "low altitudes"—underwater. Experimenting with animals, he found that nitrogen, which constitutes four-fifths of every breath we take, usually passes harmlessly in and out of the body. Under pressure, however, the gas is forced into solution in the body tissues. When pressure falls—that is, as a diver rises to the surface—the nitrogen again turns into bubbles of gas. The tiny bubbles Boyle had observed were like the bubbles of nitrogen that caused the bends in undersea workers. The amount of gas, Bert showed, varied, depending on the amount of pressure experienced and the length of exposure. A gradual return to surface pressure would allow the gas to escape naturally and avoid the agonizing consequences of decompression sickness.

Bert was the first to prove the value of decompression through controlled experiments. He was also the first to show that recompression—putting victims back under pressure when they are suffering from the bends—offers relief from the condition, as well as the possibility of controlled, safe return to surface pressure.

Some years later Professor Leonard Hill vividly confirmed Bert's theories using natural divers—frogs—as subjects. Hill placed the frogs under pressure equivalent to that more than 600 feet undersea, then rapidly brought them back to surface pressure. He quickly placed the web of one of the frog's feet under the microscope. At first the blood flowed normally; ". . . then small, dark bubbles, first one, then another, then numbers of them scurried through the vessels and drove the corpuscles before them." Hill rapidly put the frog back under pressure and soon the bubbles disappeared. The frog was not injured.

Failure to decompress properly can have crippling or even deadly consequences. James Sweeney in his book on oceanographic submersibles includes a diver's description of how the bends affected him: "On deck I took my mask off. My tender handed me a cigarette and I couldn't hold on to it. I just lost all coordination. . . . I was dizzy and I couldn't seem to maneuver." Even after treatment, the diver was partially paralyzed for several weeks.

Such devastating effects of nitrogen bubbles on the body can be avoided. The first to regularize decompression time in scientifically based tables was scientist John Haldane, "a deep-eyed Scotsman with a powerful nose, jutting chin, and a great scraggly moustache." The schedule was determined by Haldane in 1906 in a series of underwater experiments for the British Admiralty Deep Diving Committee.

Under Haldane's guidance, Guybon Damant and Andrew Catto attempted the deepest dive up to that time—210 feet—more than twice the depth reached by the Royal Navy using conventional diving procedures. Wearing heavy helmet dress and burdened with weights, the two men sank to the bottom. There Damant collected samples of air from his helmet for Haldane's experimental analysis.

On the way back to the surface, the divers carefully followed a schedule of decompression stages. At a stop at 110 feet Damant vented a small amount of air from the sample bottle to keep it from exploding as the outside pressure decreased. Both Catto and Damant emerged unharmed from the record-making excursion. It would be nearly a decade before anyone went deeper. Haldane's experiments represented a breakthrough in technology that enabled divers to plunge routinely to 200 feet and return safely.

Nitrogen under pressure causes another peculiar effect. Starting at about 100 feet underwater, the gas begins to cause a noticeable mind-bending narcotic reaction commonly known as "rapture of the deep." Jacques-Yves Cousteau described its effect on him during a deep dive in the Mediterranean Sea: "At two hundred feet I tasted the metallic flavor of compressed nitrogen and was instantaneously and severely struck with rapture. I closed my hand on the rope and stopped. My mind was jammed with conceited thoughts and antic joy. I struggled to fix my brain on reality, to attempt to name the color of the sea about me. A contest took place between navy blue, aquamarine and Prussian blue. The debate would not resolve. . . .

"I hung witless on the rope. Standing aside was a smiling jaunty man, my second self, perfectly self-contained, grinning sardonically at the wretched diver. . . .

"I sank slowly through a period of intense visions. . . . Then I went to the last board, two hundred and ninety-seven feet down. . . . I was sufficiently in control to remember that in this pressure, ten times that of the surface, any untoward physical effort was extremely dangerous. I filled my lungs slowly and signed the board. I could not write what it felt like fifty fathoms down."

ROBERT B. GOODMAN

Not until 1935 was such underwater "drunkenness" explained by physiologist-physician Dr. Albert Behnke as being caused by breathing nitrogen under pressure. By replacing the nitrogen in compressed air with helium, rapture of the deep could be avoided.

PROBLEMS OF PRESSURE, cold, and nitrogen narcosis all pale to insignificance in the minds of some at the thought of another hazard of the deep: that of dangerous sea creatures. Few animals on this planet have so run away with human imagination as have sharks, especially the great white shark.

It is difficult to convince nondivers that sharks and barracudas, sea snakes and moray eels, do not generally represent overwhelming hazards—that the "live and let live" attitude works. And it is equally difficult to convince an experienced diver that he should harbor a well-placed fear of sea creatures—with a few notable exceptions.

Al Giddings witnessed a rare shark attack and helped rescue the victim. He had just surfaced from a dive and returned to a waiting boat, anchored off the Farallon Islands near San Francisco, when he heard a loud, unearthly scream.

"I had never heard such an agonized cry before," said Al. "I jumped in and was fairly close to the commotion before I saw that the person in trouble was my business partner, LeRoy French. He called to me just before an immense triangular fin passed behind him like a monstrous gray sail. Unwittingly my face reflected what I had seen, and this, in turn, was wordlessly communicated to LeRoy just before the shark—a great white—grabbed him for the second time.

"An endless time seemed to pass before LeRoy reappeared a few feet from me, bleeding horribly. We were in a sea of red.

"I held him close and swam for the boat, certain that we would both feel the shark's jaws close on us at any moment. But we didn't see him again. Somehow we managed to reach safety. Miraculously

Fearsome monster of the deep? No—a harmless swell shark, named for its distinctive behavior. When taken from the water, the fish swells up by filling its stomach with air. Rows of teeth—awesome but minute—cannot do serious damage. Divers encounter these sharks along the coasts of California and Mexico.

JACK DRAKE / BLACK STAR

LeRoy survived, thanks to 312 stitches and a strong constitution."

Shark attacks, including LeRoy's, tend to occur when spearfishing has been on the day's agenda or when the sharks have been attacked first by humans. Early in 1980 Al and Dr. John McCosker, director of the Steinhart Aquarium in San Francisco, witnessed at close range the awesome power of great white sharks in action. The two men watched from the relative security of a cage suspended in the sea near Dangerous Reef, Australia. "There were four cruising around us at one time, attracted to the bait we set," McCosker told me. "They bit everything in sight, the bait, the boat, the cage. But I couldn't resist reaching out to touch them as they moved by—sleek, powerful, ageless—the epitome of an efficient, effective, open-sea predator." McCosker points out that while sharks, worldwide, attack perhaps 100 people a year, humans annually consume hundreds of thousands of sharks. Norway alone ships several million pounds to Europe each year. One man, Captain William Young, boasts of having killed 100,000 sharks during 60 years of hunting.

Of the more than 250 species of sharks in the world, only a few kinds have been implicated in attacks on people, and even the fearsome great whites have suffered far more at human hands—for sport, food, and display—than have humans at the teeth of the sharks.

Ron and Valerie Taylor, Australian naturalist-photographers who probably have had as much experience with great white sharks as any other divers in the world, believe that occasional attacks—and much publicity—have given sharks a distorted image. They fear that indiscriminate slaughtering may endanger the survival of these large predators. Valerie said: "Each time I am privileged to see a great white, the feeling of awe and amazement is the same. . . . Very few sharks pose a serious threat to man. Even when interfered with in the cruelest fashion (speared or hooked), sharks, the great white included, seek only to escape. . . . Sharks play a vital part in the ecology of two thirds of the world's surface. Nature ensures that the delicate balance of life in the sea is maintained. The predatory instincts of sharks help control that equilibrium."

R. EUGENIE CLARK, a diving ichthyologist who specializes in studying shark behavior, agrees. The petite, dark-eyed scientist spent many weeks observing a group of sharks in Mexican waters to try to determine why the large, streamlined creatures were behaving in a most uncharacteristic manner. Certain caves in the area were occupied by what appeared to be "sleeping sharks." The creatures remained immobile, apparently unfrightened, when divers approached them. Dr. Clark entered one of the caves with a student and found herself ". . . face-to-face with one of the sea's most deadly denizens, in the most dangerous situation possible for confronting a shark—the fish crowded, backed into a corner." Far from being terrified, Dr. Clark and her assistant, Anita George, were delighted at their ability to have such close access to a large shark in its own environment and to learn something of the creature's habits. They concluded that the sharks were attracted to the caves because the low salinity of the water there allowed the sharks to be cleaned of parasites by small fish called remoras.

Such studies of the habits of animals whose history spans 300 million years of living in the sea are invaluable to sport divers and scientists alike. The discoveries they yield may help us devise new equipment that will take us ever farther into the underwater realm.

World's largest fish, a whale shark, casts a seemingly baleful eye on remoras cruising alongside in the Red Sea. Whale sharks usually reach about 35 feet in length but pose no threat to divers. A snorkeler (bottom, right) touches down above the open mouth of one of the creatures. Of the more than 250 species of sharks, only a few attack swimmers. A horn shark (below) submits to handling off California's Santa Cruz island. If angered, the fish will sometimes bite a diver, but its small teeth render it relatively harmless. In the Tuamotus of the South Pacific, islanders have built a tank where they release captured sharks; a youngster of the islands (bottom, left) hitches a ride on a nurse shark.

JACK DRAKE / BLACK STAR; JACK McKENNEY (RIGHT)

AL GIDDINGS / SEA FILMS, INC.

RON AND VALERIE TAYLOR

"*Fierce, fearless, voracious . . . described as the world's most ferocious animal,'' says one authority about the great white shark. Responsible for many attacks on swimmers, the great white roams in most of the world's warmer oceans. Triangular, jagged teeth grow in rows; as one tooth falls out, another moves forward to replace it. When the great white seizes prey, it shakes its own body savagely. The rough-edged teeth act like a saw, neatly severing a chunk of meat. No natural enemies prey on the great white; it fears only others of its kind and—increasingly—man. Inspired by the movie* Jaws, *more and more fishermen prize the huge fish as trophies. A flourishing trade in jewelry made from sharks' teeth contributes to the killing.*

AL GIDDINGS / SEA FILMS, INC.

DAVID DOUBILET

Dozing, drugged, or just resting? Behaving contrary to established patterns, a deadly reef shark lies motionless in a cave off Mexico's Isla Mujeres (below). Mexican divers discovered the caves, where sharks unaccountably seem to sleep—contradicting the theory that the creatures need to move in order to breathe. Dr. Eugenie Clark, noted ichthyologist, studied the animals in 1973 and 1974. At left, a shark ignores two of her assistants in spite of blazing camera lights. Opposite, center: Dr. Clark flirts with danger by examining a bull shark snagged on a baited hook. Studies conducted by the divers (opposite, lower) indicate that fresh water seeping into the caves lowers the natural salinity, possibly "tranquilizing" the sharks. Also, remoras can more easily clean the sharks of parasites in the altered environment, making the caves "housekeeping" stations.

AL GIDDINGS / SEA FILMS, INC. (ABOVE); RON AND VALERIE TAYLOR (BELOW)

STEPHEN C. EARLEY

Dangers of the deep include poisonous creatures. At left, author Sylvia Earle captures a sea snake during a study off Australia. For most sea-snake venom—many times more virulent than any land snake's—no antivenin exists. At bottom, right, another of the serpents lurks beneath reef corals. Bottom, from left: The scorpionfish has spines on its dorsal fins that inject venom into its prey. The gaudy lionfish, only three inches long, drifts placidly; predators recognize its needlelike appendages as weapons. Another scorpionfish blends with pilings off a Key West beach. The puffer fish (below) harms only those that eat it: Poisonous alkaloids saturate its flesh.

SOAMES SUMMERHAYES

FREDERICK R. McCONNAUGHEY

BEN CROPP

"Gamesome and light-hearted," Herman Melville called humpback whales such as the one below. Frolicking with divers off Maui (right), the enormous creatures pose little threat—though they outweigh man hundreds of times. As they swim, humpbacks often sing, repeating long, complex sequences of resonant, throbbing melodic phrases.

SYLVIA A. EARLE (BELOW AND RIGHT)

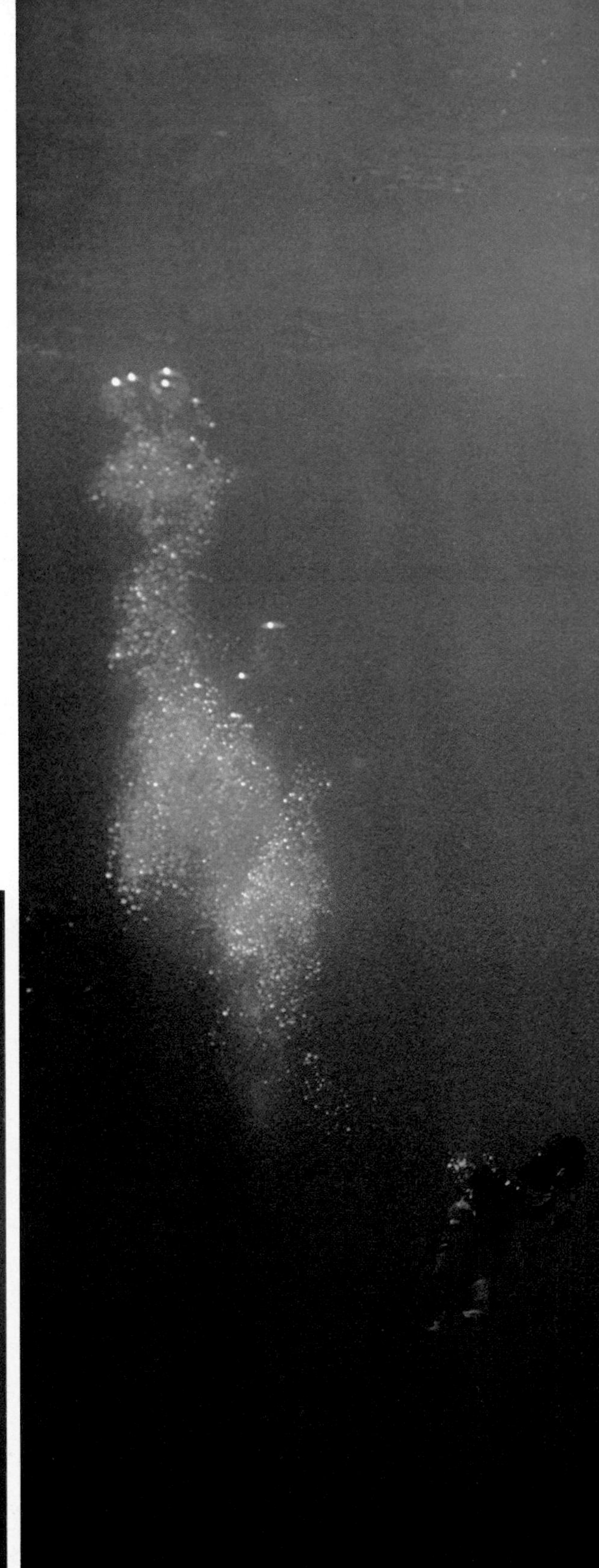

AL GIDDINGS / SEA FILMS, INC.

Good-natured friend from the sea swims with Sylvia Earle off the Bahamas (left). Before its unexplained disappearance, this wild creature—an immature spotted dolphin named Sandy—frequently swam and played with scuba divers near San Salvador. The dolphin "is the only creature who loves man for his own sake," remarked Plutarch nearly 1,900 years ago. FOLLOWING PAGES: *Curious and benign, a bottlenose dolphin surfaces off California. One of the sea's most intelligent creatures, the dolphin works easily and patiently with researchers. The Navy has trained dolphins to dive as deep as 1,500 feet. Their system of echolocation surpasses man-made sonar. A 1972 law making it illegal "to harass, hunt, capture, or kill" marine mammals without a permit has reduced the number of dolphins that drown in tuna nets each year.*

DAVID DOUBILET (FOLLOWING PAGES)

Record-breaking diver Jacques Mayol extends man's breath-holding penetration of the sea: On November 23, 1976, off the island of Elba, he reached 328 feet by holding his breath for 3 minutes, 40 seconds. At right, Mayol prepares himself by performing yoga exercises. Moments before the dive, he sits on a platform. Below: Holding a specially equipped sled, Mayol plummets bottomward in a descent that took 105 seconds. He pauses briefly at an apparatus that marks the depth, then turns toward the surface for his ascent.

BRUNO RIZZATO / MARIO BENNA

Hurrying toward the light and a blessed breath of air, Mayol ascends after his dive. Bubbles rise from eight support divers spaced along the rope that marked his descent. As cameras click and grasping hands reach out to help him up, he rests on the surface diving platform (left). Why does Mayol attempt such hazardous dives? "I think," he says, "that the human body hides

BRUNO RIZZATO / MARIO BENNA

a sort of mechanism, inherited from the past, perhaps millions of years ago. . . . When this mechanism is recognized and recovered, men will be able to work underwater at very great depths, without air tanks, like dolphins, seals, and whales. When I am down there, I'm not a man anymore. I'm a sea creature . . . a diving mammal. I belong to the water.''

Resurrected after more than three centuries on the muddy bottom of Stockholm harbor, the 17th-century warship Vasa*—once the pride of the Swedish Navy—boasts an intact hull and lower deck, complete with gun carriages (below).* Vasa *shipped water and sank on the day of her launching. Luckily, the Baltic Sea's low salinity inhibited growth of teredo worms that otherwise would have*

NATIONAL MARITIME MUSEUM OF SWEDEN

"*S*QUALUS IS DOWN off the coast of New Hampshire, depth between 200 and 400 feet. Have your divers and equipment ready to leave immediately."

"This terse command," Dr. Albert Behnke told me, "launched a sea rescue and salvage that changed diving history."

Emergency equipment of the U. S. Navy—amphibious aircraft, ships, and divers—rushed to the scene of the disaster. Help was on the way as the submarine rescue ship U.S.S. *Falcon* and its expert divers and trained officers steamed northeast from its Connecticut base. Dr. Behnke—then a Navy lieutenant himself—was among those swept into weeks of high drama. Other personnel included Dr. Charles Shilling, a medical officer who pioneered tests in submarine escape, and Comdr. A. R. McCann and Lt. Comdr. C. B. Momsen, developers of the McCann rescue chamber and a new breathing system, the Momsen lung.

"*Squalus* sank after a freak accident that flooded the submarine during trials," Behnke continued. "Twenty-six men perished immediately that day in 1939, and the fate of the 33 trapped survivors was tenuous. Their air supply was limited and subject to possible contamination from chlorine in the sealed-off battery compartment. The depth, moreover, was a matter of some concern."

At 243 feet *Squalus* was resting nearly twice as deep as the submarine *S-4* that had sunk off Massachusetts after a collision in 1927. The *S-4* crew survived the sinking but slowly suffocated while awaiting rescue that never came.

"The *S-4* catastrophe caused the Navy to develop a rescue plan," Behnke said. "Salvage crews practiced and drilled for years, learning to operate the McCann chamber and the new lung."

The temperature inside the *Squalus* fell to 45°F (8°C), with almost 100 percent humidity. The rescue chamber, never tested in an actual emergency, was successfully mated to the escape hatch of the sunken sub. Air lines were inserted into the open hatch, and fresh air from the surface revived the men inside. Cheers greeted the haggard crewmen as they surfaced aboard the *Falcon.* The chamber made four trips—within 12 hours—before the last group of survivors reached the surface. But there remained the recovery of the bodies of those who had drowned and of the submarine herself.

Behnke leaned his long, straight frame forward, and smile lines creased the corners of his strikingly blue eyes as he remembered the success of the mission: "During the next three and a half months 628 dives were made, 302 in excess of 200 feet," Behnke continued. "We began using compressed air, but that proved to be too dangerous, mostly because of nitrogen narcosis. The performance of divers

"I reached the bottom in a
state of transport. . . .
To halt and hang attached to
nothing, no lines or air pipes to
the surface, was a dream.
At night I had often had visions
of flying by extending my arms as
wings. Now I flew without wings.
(Since that first aqualung flight,
I have never had a dream
of flying.)"

Capt. Jacques-Yves Cousteau,
The Silent World

Fifty feet down off the Virgin Islands, marine zoologists from an undersea laboratory measure the effects of pollutants on delicate corals; heavy tape holds vials of chemicals. One of many technical advances since 1960, undersea quarters allow divers to live and work for months instead of minutes, decompressing only when ready to return to the surface. As inhabitants instead of visitors, they learn the secrets of the sea.

NATIONAL GEOGRAPHIC PHOTOGRAPHER
BATES LITTLEHALES

breathing air at great depths in darkness and in cold water was ineffective. The men were often confused and suffered lapses of memory and a loss of consciousness. Even though the equipment necessary for supplying helium and oxygen to deep-sea divers had not been assembled, we decided that it was time to make it happen. Air simply did not give us the depth capability we needed."

ALLAN C. FISHER, JR., NATIONAL GEOGRAPHIC STAFF

The use of oxygen combined with helium for a breathing mix was still experimental, as were the appropriate decompression tables. But there wasn't time to practice. The Navy team, including medical officer Behnke, used what they knew. Along the way they developed techniques that forever transformed diving. The key to their success was a mysterious gas, helium.

HELIUM—AN ODORLESS, TASTELESS, invisible element, the lightest next to hydrogen but without hydrogen's explosive properties—was once thought to occur only in the vapors surrounding the sun; hence the name, from the Greek word *helios.*

Eventually the gas was found on earth as well and predictably could be recovered when certain radioactive materials were boiled in sulfuric acid. Its price in 1900—$2,500 a cubic foot—reflected its rarity and also accounted for its lack of practical application until large quantities were discovered in natural gas wells in Texas. Today this magical, lighter-than-air gas sells for about 20 cents a cubic foot and fills balloons at fairs and amusement parks around the world.

The use of helium for diving was independently suggested by several scientists because of the characteristics of the elusive sun gas: Helium is chemically inert and only one-seventh as dense as nitrogen. By the early 1920s helium-oxygen mixtures were being tested by researchers with the U. S. Navy and the Bureau of Mines. Experiments on animals and later on human subjects clearly proved the advantages of the gas mixtures—which eliminated the problem of nitrogen narcosis.

In 1925 the Navy and the Bureau of Mines, which handled the supplies of helium, began deep-sea diving experiments some 300 miles from the nearest salt water—in Pittsburgh, Pennsylvania. Divers who used the new mix of oxygen and helium during the preliminary tests were able to go as deep as ever and return to surface pressure in one-fourth the decompression time, remaining clearheaded throughout. However, they noted two unexpected drawbacks: The miracle gas drained body heat much faster than air and, because of its extreme lightness, caused vocal cords to vibrate more quickly. Breathing helium, the deepest basso sounds like an excited Mickey Mouse; even frogs lose the ability to croak and render birdlike chirps instead. My children eagerly sacrifice party balloons for the hilarious pleasure of becoming instant sopranos after inhaling a few breaths of balloon helium. But high-pitched garbled sounds are no joke to divers whose survival may depend on clear messages. Despite these disadvantages, helium was generally adopted for use by commercial and Navy divers with striking results. For years they have used it in combination with other gases to work safely in the depths.

An intrepid young Swedish engineer, Arne Zetterström, was not content even with the new range opened by helium. He attempted to adapt hydrogen in its place, although that lightest gas of all is explosive unless the amount of oxygen mixed with it is very small—4 percent or less. At surface pressure such a low percentage of oxygen will not sustain human life, but at 100 feet it is adequate—and hydrogen no longer a hazard.

Zetterström worked out a method for descending to 100 feet on compressed air, then switching over to the safe hydrogen-oxygen mix. The strategy worked. On August 7, 1945, he dived an astounding 525 feet in the Baltic Sea. But, as he began decompressing, his surface tenders made an error and inadvertently pulled him through the

SUBMARINE FORCE LIBRARY AND MUSEUM, GROTON, CT.

critical 100-foot zone without the necessary stop and change-back to compressed air. The effectiveness of the hydrogen-oxygen technique was proved, but the experiment cost Zetterström's life, and only a few countries are researching the use of hydrogen today.

While some applied their genius to finding ways of going deeper, others responded to another vision—to become free in the sea, to leave behind hoses and helmets, to dive and swim as independently as a dolphin or a fish. As early as the late 1600s an Italian physicist, Giovanni Borelli, had designed a self-contained, if impractical, diving device that recycled air.

But the first real step toward success came in 1865, with an invention by two Frenchmen, Benoît Rouquayrol and Auguste Denayrouze. Their surface-supported suit had a reservoir of air on the diver's back that allowed him a few minutes of independence from the main air supply. For the first time air could be breathed when needed, instead of flowing continuously around a diver's face.

Five years later Jules Verne borrowed the idea to provide the crew of the fictional craft *Nautilus* with the means to walk about freely on the ocean floor. " 'You know as well as I do, Professor,' " Verne wrote, in a conversation between Captain Nemo and Professor Aronnax, " 'that a man can live under water, providing he carries with him a sufficient supply of breathable air. In submarine works, the workman, clad in an impervious dress, with his head in a metal helmet, receives air from above by means of forcing pumps and regulators.'

" 'That is a diving apparatus. . . .'

" 'Just so, but under these conditions the man . . . is attached

Dragged from the depths four months after sinking on May 23, 1939, the U.S.S. Squalus *churns the waters off New Hampshire as four support ships stand by to help (above). The nine-month-old vessel flooded on a test dive and sank to 243 feet, trapping 59 men aboard. Although 26 sailors drowned, a new diving bell, similar to the rescue chamber shown opposite, enabled the Navy to save 33 men imprisoned in the forward section. Four times rescuers lowered the steel bell on a cable to the submarine's escape hatch, clamped it there, and brought up trapped crew members. A winch reeled the device to a waiting ship, where the men emerged.*

to the pump which sends him air through an india-rubber tube, and if we were obliged to be thus held to the *Nautilus*, we could not go far.'

" 'And the means of getting free? . . .'

" 'It is to use the Rouquayrol apparatus. . . .' "

U. S. NAVY

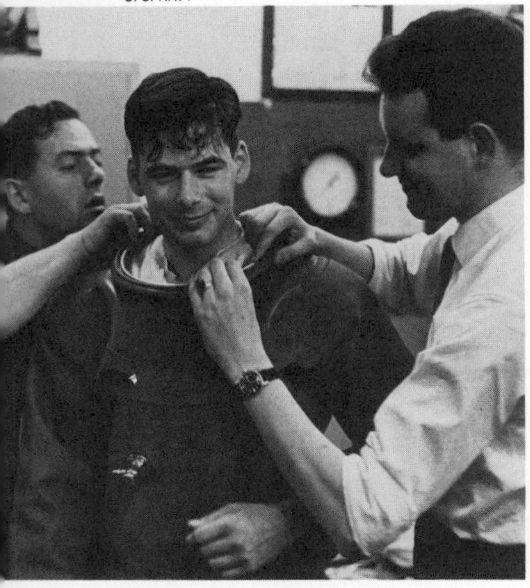

Suiting up, Swiss mathematician and deep-diving pioneer Hannes Keller prepares for a plunge to a simulated depth of 700 feet at the U. S. Navy's Washington, D. C., Weapons Plant in 1961. Breathing his special mixture of gases, Keller executed several successful tests like this one before diving with a partner to 1,000 feet off the coast of California in 1962. He set a depth record then, but his companion, Peter Small, died before surfacing, and a support diver drowned.

IN THE NONFICTIONAL WORLD, a successful open-circuit system of scuba—self-contained underwater breathing apparatus—was patented and used in 1918 by a Japanese inventor, Ohgushi. Like the others, his system employed compressed air in tanks, but it differed in that air flow was triggered by biting on a special mouthpiece.

The direct precursor of modern scuba appeared in 1933 when Yves Le Prieur devised a diving system that released air into a diver's face mask from a bottle of compressed air carried on his chest.

But, justifiably, Jacques-Yves Cousteau and Emile Gagnan are credited with the first practical scuba to gain wide acceptance, the Aqua-Lung. Cousteau's vision is described in *The Silent World:* "We were dreaming about a self-contained compressed-air lung. Instead of Le Prieur's hand valve, I wanted an automatic device that would release air to the diver without his thinking about it, something like the demand system used in the oxygen masks of high-altitude fliers. I went to Paris to find an engineer who would know what I was talking about. I had the luck to meet Emile Gagnan, an expert on industrial-gas equipment for a huge international corporation. It was December, 1942, when I outlined my demands to Emile. He nodded encouragingly and interrupted. 'Something like this?' he asked and handed me a small bakelite mechanism. 'It is a demand valve I have been working on to feed cooking gas automatically into the motors of automobiles.'

"In a few weeks," continued Cousteau, "we finished our first automatic regulator."

I was among those who read and eagerly reread *The Silent World* and the October 1952 NATIONAL GEOGRAPHIC article in which Captain Cousteau shared his incredible experiences as a "fish man." He said then, "The best way to observe fish is to become a fish. And the best way to become a fish—or a reasonable facsimile thereof—is to don an underwater breathing device called the Aqualung. The Aqualung frees a man to glide, unhurried and unharmed, fathoms deep beneath the sea. It permits him to skim face down through the water, roll over, or loll on his side, propelled along by flippered feet. . . . In shallow water or in deep, he feels its weight upon him no more than do the fish that flicker shyly past him."

The way was open for a new era of man-in-the-sea to begin.

When the Russian satellite Sputnik arched into space in 1957, it gave a sudden new dimension to human aspirations. A *New York Times* editorial summed up the expectations of the time: "The creature who descended from a tree or crawled out of a cave a few thousand years ago is now on the eve of incredible journeys." The same engineering ingenuity that devised ways to go farther skyward and stay longer aloft was directed toward exploration of the oceans. The early 1960s marked a climax of these two crescendos, a simultaneous ascent and descent into outer and "inner" space.

In 1960 Swiss inventor Jacques Piccard and Lt. Don Walsh, USN, plunged nearly seven miles aboard the bathyscaph *Trieste* into the Mariana Trench, the deepest place known in the sea. They provided a new perspective on the earth from the ocean depths during the same era in which cosmonaut Yuri Gagarin became the first human to orbit the earth and view the world from a perch in space. And in 1962, the year John Glenn dramatically circled earth in *Friendship 7,* a young Swiss mathematician, Hannes Keller, and his companion, Peter Small, took a step in another direction. They descended 1,000 feet into the sea off California, while breathing a mix of gases developed by Keller. The dive set a depth record but caused Small's death.

J. BAYLOR ROBERTS

"To earthbound man he gave the key to the silent world": Flanked by President John F. Kennedy and Dr. Melville Bell Grosvenor, then Editor (now Editor Emeritus) of NATIONAL GEOGRAPHIC, *Jacques-Yves Cousteau displays the inscription on the Society's Gold Medal, presented to him at the White House in 1961. Designs on the award—Aqua-Lung divers, the research ship* Calypso, *and his diving saucer—mark highlights of Cousteau's career. For 15 years the Society cosponsored his undersea explorations.*

The mixture was another response to the problem of breathing gases under pressure. Each gas in a mix exerts independent pressure—called partial pressure—on the lungs. The partial pressure of oxygen, as well as of gases such as nitrogen and helium, is nearly doubled with every atmosphere of pressure added. The characteristics of each gas create separate physiological dangers that limit how deep a diver can go with a given breathing mixture.

For example, oxygen—normally thought to be a friend—is a foe when breathed at depths greater than about 30 feet. Its effects can be dramatic, convulsive, even fatal. As Zetterström proved, hydrogen, normally regarded as dangerous because of its highly explosive nature, is the most diffusible of all gases. When combined with a small amount of oxygen, it may give divers even deeper diving capability than helium. Nitrogen, harmless at the surface, causes narcosis in the deep.

Keller took all this into consideration and developed a breathing mixture of three gases—nitrogen, helium, and oxygen—for his unprecedented short-decompression deep dives. He later revealed the proportions of the gases in his formula to the Navy.

Meanwhile, the dream of living and working underwater, not as visitors but as long-term undersea residents, began to be realized—largely because of the purposeful genius of three men: Capt. George F. Bond, USN, affectionately known to his friends as "Papa Topside," Capt. Jacques-Yves Cousteau, and engineer-inventor Edwin A. Link.

Bond looks out on the world from under shaggy eyebrows, with a smile lighting up his eyes or creasing his sun-browned face much of the time. His proposal was to extend man's undersea work capability with respect to both duration and depth. Bond pointed out that once a diver's body was saturated with compressed gas—when tissues had absorbed all they could absorb and equilibrium was reached—the decompression time would be the same whether the diver stayed underwater for a matter of hours, days, weeks, or even months. The amount of time necessary for decompression depended on the depth of the dive and on the gases breathed.

To live underwater, Bond reasoned, it would be necessary to bring air pressure inside a chamber—an underwater house—into equilibrium with the water pressure outside. Then divers could swim freely in and out of their warm, dry apartment into the surrounding sea. Verne had imagined such things in *Twenty Thousand Leagues Under*

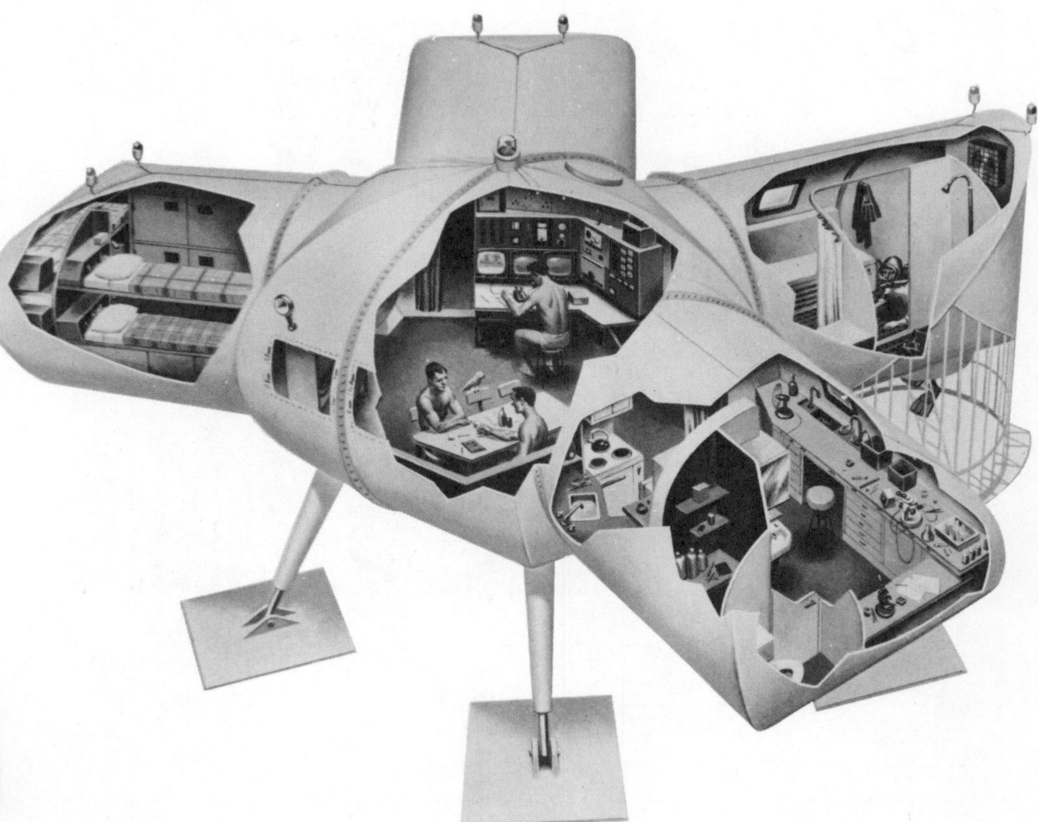

Comfortable quarters, Starfish House (right) lodges five Conshelf Two divers 36 feet deep in the Red Sea. For a month in 1963 this second stage of Cousteau's continental shelf project tested man's ability to withstand long, deep dives. Radiating from a control center and living-dining area, the habitat's arms contain a kitchen, laboratory, ready room, and bedrooms. A sharkproof grille protects the exit. First submersible to operate from a base on the ocean floor,

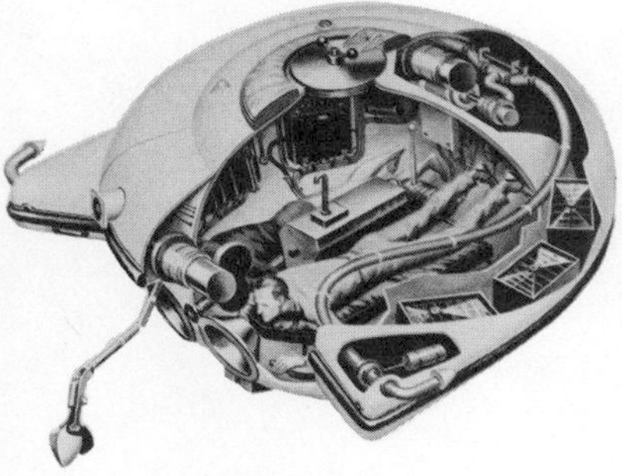

Cousteau's diving saucer (above) added another dimension to the Conshelf experiment.

PAINTINGS BY DAVIS MELTZER

the Sea, and Bond provided the know-how to bring the concept to life.

Five years of data painstakingly produced in U. S. Navy laboratories preceded the first successful open-water saturation dives. According to Bond, "In late 1959 Jacques Cousteau and I appeared together on an undersea program in New York. Afterwards, the late undersea chronicler Jim Dugan invited me to his apartment for further talks with Cousteau. We discussed my concept of saturation diving. In fact, we talked until breakfast!"

About a year later Bond successfully exposed human volunteers in a pressure chamber to a simulated depth of 200 feet for 14 days and communicated his results to Cousteau and to Edwin Link. The U. S. Navy did not expect to undertake open-sea tests for at least five years, but with encouragement from Bond, Cousteau and Link proceeded on independent saturation-diving programs.

The world's first undersea station, at 60 feet, was occupied for 14 hours (including six hours for decompression) in 1962 by its designer, Edwin Link. Although 58 and "retired" at the time, Link surprised no one with his participation. With his long track record of personal involvement in varied aviation and submarine creations, it was natural for him to try submerging himself in the new chamber before relinquishing space to Robert Sténuit for a deeper (200-foot), longer (24-hour) stay a month later. Sténuit became the first open-sea saturation diver. His total decompression time was 66 hours (including 58 hours aboard ship)—a long trade-off for 24 hours in the sea—but had he stayed for months, not hours, the decompression time would still have been the same.

The same issue of NATIONAL GEOGRAPHIC that described the innovative dive by Sténuit and Link also brought news from Venus. The spacecraft Mariner II had reached the lemon-yellow clouds of earth's sister planet and brought back data showing that Venus was an inhospitable place by human standards, with an average surface temperature of 800°F (426°C). No life as we know it could exist there.

In striking contrast, reports from the aquanauts were crowded with enthusiastic accounts of encounters with amazing, bizarre creatures. Some had not been seen before by human eyes, but even familiar fishes assumed a new character now that it was possible to meet them on their own terms, day and night.

A FEW DAYS AFTER Sténuit opened the decompression chamber and stepped out, Jacques Cousteau looked down on a large yellow chamber in 33 feet of water in the Mediterranean Sea near Marseille. Conshelf One was under way, and the dwelling was about to receive its two residents, Albert Falco and Claude Wesly, for a week of underwater living that Cousteau called a "logistical experiment more than a physiological one."

Comparing undersea living to going into the ocean as part of a submarine crew, Cousteau observed, "The submariner was not adapting to the sea, but holding it back with blind steel plates . . . an in-patient, forbidden to look at his world except through a periscope, while our men were living in the open water. . . ." He described the underwater station as an enormous Aqua-Lung into which Falco and Wesly retreated. From it they ventured forth as explorers, to observe the sea close at hand.

Conshelf Two, in June 1963, was Cousteau's effort to establish the first human colony on the seafloor. The main settlement, Starfish House, was anchored 36 feet down in the Red Sea. This strange golden structure kept five men healthy and working for a month. Four arms radiated from the center of Starfish House: One contained kitchen, laboratory, darkroom, and toilet; in another a sharkproof grille guarded the habitat's entrance and admitted divers to the ready room and shower. The two remaining arms held four bunks each. The divers relaxed in the center section—combination living room and dining room—in the company of a parrot, Claude. If the air grew noxious, the parrot would die. Simone Cousteau, Captain Cousteau's wife, arrived for the final week of the experiment. Some distance away, in 90 feet of water, was a second station, Deep Cabin, where two men lived and from which they made excursions to depths as great as 165 feet. Other parts of the habitat included Cousteau's diving saucer and its domed seafloor hangar.

The excitement of this kind of experiment is best captured by those who were a part of the action. In a NATIONAL GEOGRAPHIC account Ed Link said of his next venture in undersea habitats: "Call it the deepest long dive. Call it the longest deep dive. Both definitions describe our goal beneath the bright water of the Bahamas . . . we had come to put two men a long way down for a long time—more than 400 feet for more than 48 hours."

The two men, veteran diver Robert Sténuit and marine biologist Jon Lindbergh, were about to enter the great yellow and black "submersible, portable, inflatable dwelling"—SPID. Link's newly designed breathing system would enable them to recycle an oxygen and helium mix from which carbon dioxide would be removed chemically. The men would be on their own, beyond the reach of surface divers,

U. S. NAVY

Astronaut-turned-aquanaut: In 1968 the senior member of the Navy's Sealab III project, M. Scott Carpenter, uses a model to describe the underwater habitat, shelved when one of the divers died. Veteran of the earth-orbiting Mercury 7 mission, Carpenter had led two of the three teams of the 1965 Sealab II experiment. Their total stay at 205 feet amounted to 45 days.

After two days at a depth of 432 feet, Jon Lindbergh, left, and Robert Sténuit, right, answer questions for Dr. Joseph MacInnis in a shipboard decompression tank. With pressure in the chamber equivalent to that of 30 feet of water, Dr. MacInnis could enter safely, but Lindbergh and Sténuit could not emerge without completing 92 hours of decompression after this 1964 Man-in-Sea test. The two men could have stayed underwater longer without increasing their time in the chamber because the human body will absorb only a certain volume of inert gas at a given depth. After about 24 hours, the tissues become saturated. Thus, "saturation diving" eliminates many long decompressions, increasing man's ability to live and work in the sea.

N.G.S. PHOTOGRAPHER BATES LITTLEHALES

sustaining 200 pounds of pressure on every square inch of their bodies.

Sténuit remarked during trials, "An ingenious idea, this undersea tent, but . . . rubber? Was that not rather fragile? I felt like a goldfish in a carry-home bag." But he shared Link's vision of the possibilities SPID provided. "I knew what Ed had in mind," he said. "To live, eat, sleep, and work in depths so far unreachable by free divers, and in so doing, to take a long step toward the conquest of the continental shelf. To me it was the most extraordinary adventure of which a diver might dream. My answer was ready when Ed asked me if I would like to spend a deep-down day or two in his shiny new can."

The moment of truth came one sunny day in June 1964. Sténuit and Lindbergh rode 432 feet to the seafloor in a special diving chamber and swam into SPID for a two-day sojourn in the sea. At the end of the experiment, four days of decompression were needed before they could rejoin their companions at the surface.

Sténuit describes the experience: "I have returned from a strange journey in an alien world. . . . Living in the depths, I have become in certain ways a creature of those depths, adapted to their pressures. Now the human environment is temporarily intolerable to me. I need pressure. Without it, the gas my tissues have absorbed would turn to bubbles. And so I must wait inside this lifesaving prison of a decompression tank until I have been slowly weaned from pressure and made once more fit to live on earth."

TWO YEARS OF exploratory dives by Link and Cousteau served as prelude to the U. S. Navy's first undersea station in July 1964. Under the supervision of Captain Bond, the Navy staged the first open-sea saturation dive, Sealab I, near Bermuda. For 11 days four men breathed a mixture of helium and oxygen at a depth of 193 feet. Their success inspired a continuing saturation-diving program for the Navy, as well as Sealab II the following year, again headed by Bond. Sealab II enabled 28 divers in three teams to explore the cold waters near La Jolla, California, while occupying an undersea laboratory for 15-day periods. Their 45-day total stay represented the longest manned undersea experiment up to that time.

One of the team members had a built-in wet suit, permanent flippers, and a persistent smile. He did not bother with the helium-oxygen mix used by the other aquanauts. "Tuffy," a trained dolphin, impressed divers with regular delivery service. Once Tuffy made seven 200-foot excursions from surface to Sealab within 20 minutes!

Meanwhile, Cousteau was launching Conshelf Three, designed for living and working in 328 feet of water in the Mediterranean off the southern coast of France. The habitat proved that subsea stations were practical for oil exploration and recovery. As amazed oil engineers onshore watched television monitors, the oceanauts—as the French undersea explorers were known—attached a 400-pound repair assembly to a wellhead more quickly than could land crews, who lacked the advantage of buoyancy.

Although their days were characterized by long hours of labor in numbing cold, dark waters, the six Conshelf Three oceanauts were clearly at home in the sea. They intended to stay for two weeks, but bad weather robbed them of eight days' work. Cousteau asked if they would be willing to stay a little longer. Their response, sent on an Electrowriter to those above, reflects a common attitude of enthusiasm in those who have lived and worked in the sea: "To the Surface People: Have mercy on us, poor little *(Continued on page 139)*

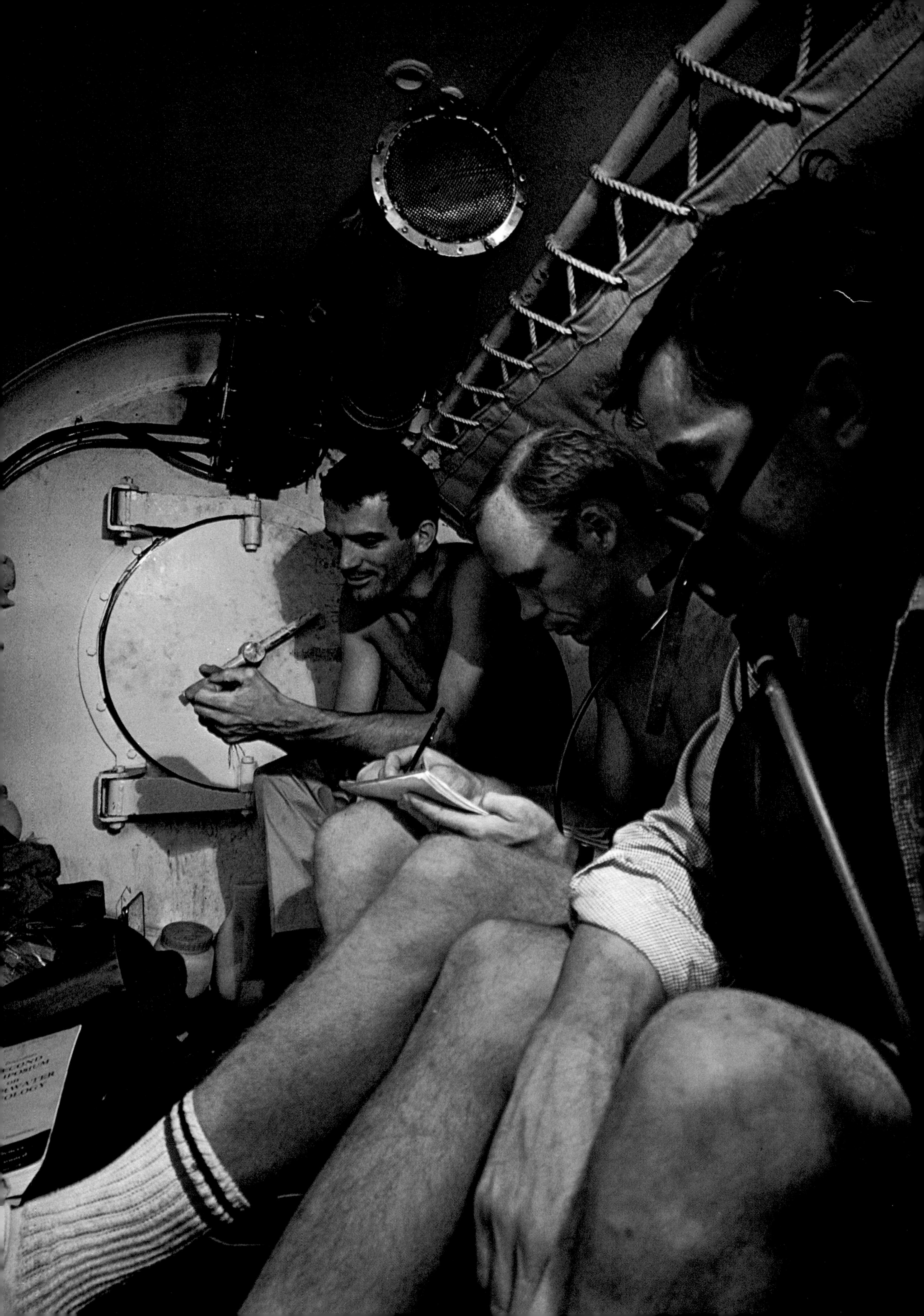

Off the Mediterranean coast of France, Edwin A. Link's submersible decompression chamber, equipped with air locks and airtight hatches, swings from ship to sea in the first stage of his Man-in-Sea Project in 1962. Robert Sténuit winched himself down 200 feet to spend a day in the two-room aluminum cylinder. Breathing a mix of oxygen and helium to avoid the narcotic effect of nitrogen at great depths, Sténuit proved that man could withstand the tremendous pressures of the deep. Despite his readiness to remain below, crewmen, under Link's direction, abruptly hauled him to the surface (above, right) after high winds sank a launch bringing vital helium supplies. For 58 hours Sténuit decompressed aboard ship in his 11-foot-long diving tank. Through a porthole, observers could view his physical reactions, read the pressure gauge above him, and even watch as he enjoyed hot coffee (right). Inventor of a

THOMAS J. ABERCROMBIE, NATIONAL GEOGRAPHIC STAFF (BOTTOM RIGHT); N.G.S. PHOTOGRAPHER BATES LITTLEHALES

famous flight simulator, Edwin Link (far right), shown on his research ship Sea Diver, *turned to undersea exploration upon his retirement. Convinced that man could live and work deep in the sea, he devised the revolutionary equipment used in Man-in-Sea.*

N.G.S. PHOTOGRAPHER BATES LITTLEHALES (ABOVE AND RIGHT)

*Submersible, portable, inflatable dwelling—*SPID*—undergoes tests off the Bahamas (right) in preparation for the second phase of Man-in-Sea, the two-day Sténuit-Lindbergh plunge to 432 feet in 1964. Divers entered the four-by-eight-foot dwelling through a passage on its underside; pressure within the shelter kept out water. Tied to the* Sea Diver, SPID *floats before the dive (above), while a crewman checks the chamber. Cramped and cold in their rubber home, the two men huddle among the pressure gauges that ensure their safety.*

ROBERT STENUIT

oceanauts in the womb of the Immense Sea. Bring us back up . . . as late as possible."

The mission had its symbolic moments as well: Team leader André Laban conversed by telephone with Sealab II aquanauts 6,000 miles away. And in an exchange characteristic of the network of exploration, astronaut Gordon Cooper, then orbiting 150 miles above the sea in his Gemini 5 spacecraft, sent greetings by way of a special radio link with Comdr. M. Scott Carpenter, USN, leader of the first two Sealab II teams then submerged 205 feet below the sea surface!

Spurred by an incredible record of successes and no major mishaps, one of the most ambitious undersea projects ever conceived—the U. S. Navy's Sealab III—got under way in February 1969. About 50 divers were trained to live and conduct scientific experiments at a depth of 620 feet, making excursions to a maximum depth of 1,000 feet. In February the spacious and sophisticated habitat was lowered to the ocean floor off San Clemente, California. Four divers were sent down to repair a helium leak. During the maneuver one of the divers died, apparently of carbon dioxide asphyxiation. Concern for the safety of other divers caused the Navy to terminate seafloor habitat programs, but studies of saturation diving continued.

Kinsmen and comrades, Jacques Cousteau and his son Philippe confer at a Monte Carlo dock beside Conshelf Three—the underwater station they designed and tested. Captain Cousteau instilled in his son his own love and understanding of the world beneath the waves. Together they probed the depths until Philippe's death in a plane crash in June 1979.

N.G.S. PHOTOGRAPHER BATES LITTLEHALES

THE WINTER OF 1969 also marked the beginning of the Tektite Project, sponsored by the U. S. Navy, the National Aeronautics and Space Administration, and the Department of the Interior. It seemed logical that tektites, small glassy meteoritic nodules found on the ocean floors as well as on land, should lend their name to a program that would reflect the nation's parallel concern for exploration of space and sea.

This project held special interest for NASA. Using a 24-hour monitoring system and special psychological tests, researchers gathered information that helped to determine how individuals would work in isolation in a hostile environment. Officials of NASA hoped this would help in preselecting individuals for participation in space flights of long duration or for work in isolation elsewhere.

Five hundred million people would peer intently at television screens a few months later to watch as the shadowy figure of Neil Armstrong placed the first footprints on the moon. In Great Lameshur Bay, off St. John in the U. S. Virgin Islands, the first Tektite explorers had been the focus of little publicity. The most numerous witnesses had been fish when four men stepped into the sea and swam to an underwater outpost on a different frontier.

In the record-making two-month stay that followed, the four scientist-aquanauts made geological and biological studies and were themselves the object of intensive study. Daily they responded to medical inquiries, took blood samples, swabbed their skin for bacteria, and underwent intensive medical examinations that provided information for many dives that followed. During their stay the scientists breathed a mixture of nitrogen and oxygen, since they worked in depths shallower than 100 feet.

One year later I stepped into the sea close to the place where the first four occupants of the Tektite habitat had made their historic entry. The blue water closed over my head. As I flippered with my four companions toward our home-on-the-reef for the next two weeks, I knew that the experience of living 50 feet down would change my understanding of the ocean world.

I already had spent more than a thousand hours underwater,

using an assortment of diving equipment: helmets, scuba, and even the remarkable *Deep Diver* submersible, designed to allow divers to swim in and out of the craft on the sea bottom. But I had never before had an opportunity to have unlimited time in the sea, day and night, or the prolonged view of an undersea resident.

PAINTING BY PIERRE MION

On the floor of the Caribbean Sea, Tektite (above) provides laboratory and living quarters for teams of five scientists. In this artist's recreation one diver enters the habitat as another dries off near the open hatch; a tunnel leads to the control center and living area. The 1969-70 Tektite missions, a joint project of government, universities, and industry, emphasized the value of saturation diving in deep-sea research and provided information to the space program.

As a botanist on land I take for granted the ability to visit the fern patch of my choice and stay as long as I like. Underwater my visits are necessarily stopwatch affairs, with severe penalties in decompression time or health hazards if I overstay by even a few minutes. When I received the invitation to be part of Tektite II, headed by program manager Dr. James Miller, it didn't take me long to respond!

The continued interest of NASA in behavioral studies of the aquanauts of Tektite II included routine observations of the team I was asked to lead, although we were all women and there was no indication that NASA was ready to allow women to be astronauts. We found it rather easy to ignore the watchers who were observing our day-in, day-out activity inside the habitat. Sometimes, though, it was irresistible to tune our television monitors so we could watch the watchers watching us . . . or to indulge in the ultimate refinement of tuning the system so we could watch ourselves watch the watchers, while they, in turn, watched us. Fish, meanwhile, peered in through the portholes.

Occupants of other undersea dwellings sized up the comfortable four-room quarters in Great Lameshur Bay and dubbed this particular laboratory-home the "Tektite Hilton." Rugs softened our footsteps, bright colors lightened the rooms, and a spacious freezer held prepared meals. A clothes drier, hot-water shower, refrigerator, stove, tape deck, television set, shelves of books, and comfortable bunks created an intentional atmosphere of hominess. The concessions to comfort proved not to be luxuries but rather were conducive to good performance during prolonged isolation. Microscopes, analytic equipment, and access to the greatest marine laboratory in the world—the sea itself—provided an opportunity over a period of seven months for the

50 Tektite II scientist-aquanauts to observe the sea as no one had before. The emphasis was no longer on proving saturation-diving techniques but on using them. The advantages of increased time in depths to 70 feet—a full eight hours or more in the sea each day, if we chose—soon became evident. But the most significant difference between in-and-out visits from the surface and living underwater came with a changed viewpoint. We became reef residents.

Just as home on land is not simply the place I return to each evening, neither was the Tektite habitat. Rather, I soon came to feel that the entire reef was my backyard and an important part of home. I began to get acquainted with my aquatic neighbors. Fish were not random gray angels and queen triggerfish, but *that* group of five gray angels living at the edge of the reef and *that* queen triggerfish resting at night in a crevice beside the long tube sponges.

My four companions and I—and the 45 other aquanauts who occupied Tektite—used either double scuba tanks or a sophisticated new rebreather system that was comparable to the life-support packs worn by Apollo astronauts when they went moon-walking. The new equipment not only provided several hours of continuous use, but also was delightfully bubbleless and thus silent.

John Vanderwalker, veteran of Tektite I and II, was vividly impressed with the effects of silent breathing: "As I prowl the reef with my partner Ian Koblick, I can recognize familiar sounds—the low-pitched grunt of the Nassau grouper, the frying-pan crackle of snapping shrimp, the staccato sound of squirrelfish. Other noises rise and fade in the background, a muted symphony of whistles, clicks, and flutters that orchestrate everyday life and death in the sea. It is not . . . a notably silent world."

I agreed as I watched—and heard—a blue parrotfish noisily grazing on algae from a piece of limestone with a vigor and tempo that made me hungry for a crisp apple.

THOMAS B. POWELL III, NATIONAL GEOGRAPHIC STAFF

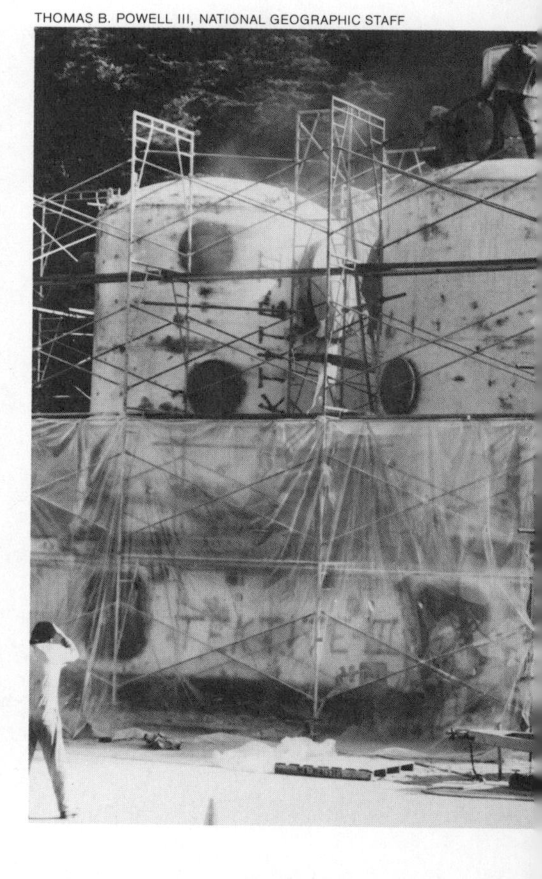

Now privately owned, Tektite will serve as a training center for underwater students off the coast of California. Refurbishing the habitat prior to installation, workmen sandblast the shell.

NIGHT WAS AS APPEALING underwater as day. I had dived after dark before, but never with the ease provided by an undersea habitat. To stay inside proved impossible when five tarpon, large silver-scaled fish with luminous eyes, arched and turned right by the window. Every night we would slip through the round hole in the floor of the room called the wet lab and into the sea beyond. We carried lights but did not always use them. In clear water, even at 70 feet, a bright moon and a sky full of stars gave enough light for us to move freely without bumping into large objects. It was also possible to see the living light of bioluminescence, to observe the blue glow characteristic of hundreds of small sea organisms.

As I swam one evening with habitat engineer Peggy Ann Lucas, we left a glowing wake as our flippers propelled us along a reef face. I was reminded of a night years before in Puerto Rico's famous Phosphorescent Bay, where brilliant one-cell dinoflagellates covered moving fish—and people—with flashing blue sparkles. Repeatedly I would dive down, then lean back and allow myself to surface face-up, enjoying the between-worlds feeling that came as the blue stars of light merged against my faceplate with silver points in the distant constellations in the Milky Way.

As an undersea dweller, I could not surface—not yet. For those of us living in Tektite, breathing a mixture of 91 percent nitrogen and 9 percent oxygen, more than 20 hours of decompression awaited us on our return to a terrestrial atmosphere. Yet I bore no feeling of resentment. Decompression time was more like the hours spent on a transcontinental—or a transatlantic—plane, time necessary to get from one world to another.

While the aquanauts of Tektite II lived in the Caribbean Sea, six other aquanauts set a new depth record for underwater living in the clear waters off Hawaii. For six days in June 1970 a strange-looking yellow habitat named Aegir, for the Norse sea god, provided an undersea haven in 520 feet of water, nearly 100 feet deeper than any free divers had dwelt before. One of the inhabitants recalled later, "I felt so free; I flew in the water like a bird in air."

Silvery air bubbles mark the trail of fishermen from Conshelf Two—Cousteau's underwater village in the Red Sea. So they would not disturb animals near the colony, the five men who inhabited Starfish House for a month in 1963 swam a short distance from their shelter to capture food. The reflective fabric of their aluminized diving suits distinguished the oceanauts from black-suited support divers.

ROBERT B. GOODMAN

FIFTY UNDERWATER HABITATS are described in the 1975 edition of the diving manual prepared by the National Oceanic and Atmospheric Administration. Some are simple, shallow-water structures that serve as underwater camping facilities, such as Subigloo, developed by undersea medical pioneer Dr. Joseph MacInnis for use as a way station under Arctic ice. Others are highly sophisticated habitats, such as the German Helgoland and the Soviet Chernomor, used for geological and biological studies.

Most of the undersea habitats developed for scientific or experimental use were retired after short programs. One notable exception is a shallow-water workhorse called Hydro-Lab. Designed and built by Perry Oceanographics, Hydro-Lab was first put to use in 1968, 50 feet down just offshore from Riviera Beach, Florida. Robert Wicklund, now Sen. Lowell Weicker's staff aide in charge of ocean legislation, began a long-range program involving the habitat near Freeport, Grand Bahama, in 1971. For more than four years Hydro-Lab was a working base of operations for hundreds of scientists and others who needed or wanted prolonged access to the sea. The Tektite Project had made a point of separating men and women, but Hydro-Lab, from the beginning, established a policy of coeducational teams. Of the 343 Hydro-Lab aquanauts, 37 were women.

In 1977 Hydro-Lab, renovated and renamed NULS I, for National Undersea Laboratory System, began sheltering a new series of scientific missions near the U. S. Virgin Islands. It is at present the only nationally supported underwater facility in the United States, but plans are under way for facilities for saturation diving. Possible sponsors include the University of Southern California, the University of Hawaii, and a consortium of southeastern universities headed by the University of North Carolina. And in the waters off San Francisco, Tektite is again being readied for use under private auspices.

Commercial diving technology proceeded through the 1970s in giant leaps, with occasional saturation dives to 1,000 feet in the toughest, coldest circumstances imaginable—in the North Sea. Although most industrial diving takes place in less than 300 feet of water, some brief excursions are made to 500 feet, and deeper saturation dives sometimes last for six weeks or longer.

Beyond 1,000 feet all saturation-diving techniques become experimental. Unshielded against outside pressure, the human body cannot endure the stress of such depths, any more than it can travel through the inhospitable atmosphere of space without a great deal of engineering ingenuity.

How to dive deeper and remain longer? A few centuries ago the limiting factor was how long one could hold a breath. A few decades ago to dive beyond 300 feet seemed impossible. Perhaps now the only limit is in our imagination.

ROBERT B. GOODMAN

Equipped with the Aqua-Lung he helped design and holding a camera, Captain Cousteau (right) explores the terrain near Conshelf Two. Another oceanaut films the landing of Starfish House (below, right). Home for two men for a week, Deep Cabin (left)

plummets downward 90 feet as divers set up the habitat. Breathing compressed air and helium, the oceanauts who lived in Deep Cabin worked at depths to 165 feet. Another Conshelf component, the diving saucer, slips into the onion-shaped dome of its hangar.

NATIONAL GEOGRAPHIC PHOTOGRAPHER BATES LITTLEHALES

In Nice harbor, Philippe and Jacques Cousteau (left) don wet suits before testing Conshelf Three in 1965. Suspended from cables, this habitat, a checkered sphere 18 feet in diameter, awaits its first underwater trial (below). Reducing their dependence on surface support, the Conshelf Three team proved that man can work as well as survive on the continental shelf. Braving 55°F (13°C) water, the oceanauts replace a 400-pound valve on a simulated oil well in 45 minutes, half the time it takes land crews to do the same task—showing that divers can operate heavy equipment in the depths.
PRECEDING PAGES: *Lights glowing in the chill waters, Conshelf Three nestles 328 feet down in the Mediterranean, housing six oceanauts for three weeks. Cousteau's diving saucer scuttles past in the gloom.*

ANDRE LABAN (PRECEDING PAGES); PHILIPPE COUSTEAU (OPPOSITE)

WALLY JENKINS

RON CHURCH / P. Q.–WOODFIN CAMP AND ASSOC.

BERNIE CAMPOLI (BOTH BELOW)

Father of saturation diving and supervisor of Sealab—the U. S. Navy's underwater habitat experiment—Capt. George F. Bond (left) monitors the shipboard tank where four Sealab I divers decompress after 11 days underwater. The 1964 project completed, a crane hoists the 40-foot-long chamber (below, left) from a depth of 193 feet off Bermuda. At right, divers prepare the larger and more complex Sealab II for submersion off California in the fall of 1965. At 205 feet below it sheltered three 10-man teams in 48°F (9°C) water.

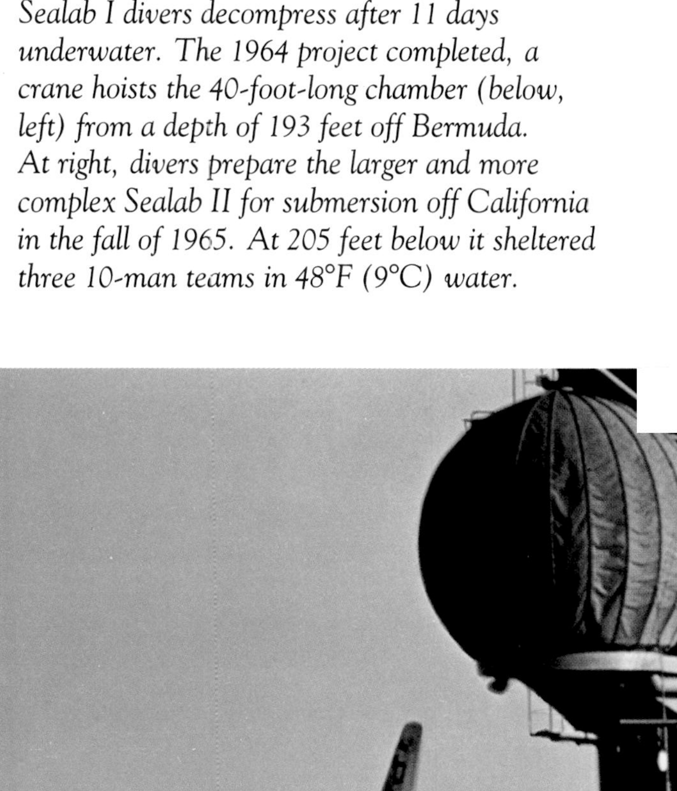

Bright yellow shell of Sealab III dominates the Naval Shipyard in Long Beach, California. The Navy shelved the 1969 project when one man died during an early test dive. Since then interest in seafloor habitats has shifted to programs involving submarine rescue.

Turquoise shallows strewn with reefs drop to sapphire depths off the Bahamas, locale of Hydro-Lab in 1971. Situated near a coral head, the 16-foot cylinder (left) sheltered four divers at once. Its liquid door, an open hatch on the underside, served as entrance and exit (below, left). Bought by the U. S. government and moved to the Virgin Islands, Hydro-Lab—renamed NULS I *—still offers divers striking close-ups of marine life, such as a group of horse-eye jacks (below, right).*

STEPHEN C. EARLEY (ABOVE); SYLVIA A. EARLE

From their twin-towered laboratory-home on the floor of the Caribbean's Great Lameshur Bay, Tektite scientists (left) studying coral respiration glide toward gnarled fingers of a sea whip. The project took its name from glassy nodules of meteoritic matter found on land and on the ocean floor. Observation of the scientists working in isolation provided data applicable to space journeys, while the experiment enabled researchers to survey the sea frontier. For seven months in 1970 a series of Tektite II teams probed the dynamics of spiny lobster populations, the vulnerability of reefs, and sonic communication among fishes. Tentacles extended, a squid (top, right) jets through azure waters near the lab. Its complex eyes and its speed make it an efficient predator. A peacock flounder 12 inches long (above) blends into the sandy floor of the bay. Another bottom dweller, a tiny octopus (right), hides in a discarded mollusk shell.

N.G.S. PHOTOGRAPHER BATES LITTLEHALES

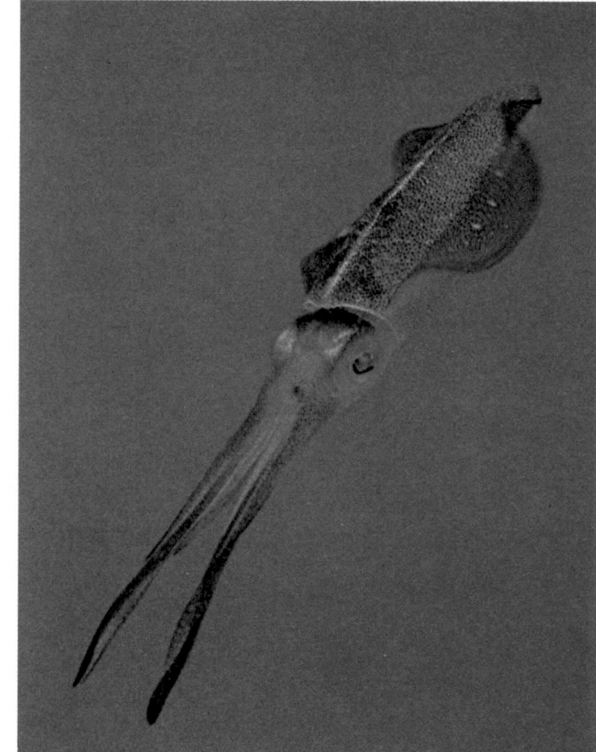

NATIONAL GEOGRAPHIC PHOTOGRAPHER BATES LITTLEHALES

PRECEDING PAGES: *White sands drift around divers from an all-woman Tektite II team in 1970. Author Sylvia Earle (foreground, middle), with the assistance of marine biologist Alina Szmant, collects plants for cataloging; ecologist Ann Hurley takes notes.*

N.G.S. PHOTOGRAPHER BATES LITTLEHALES (PRECEDING PAGES)

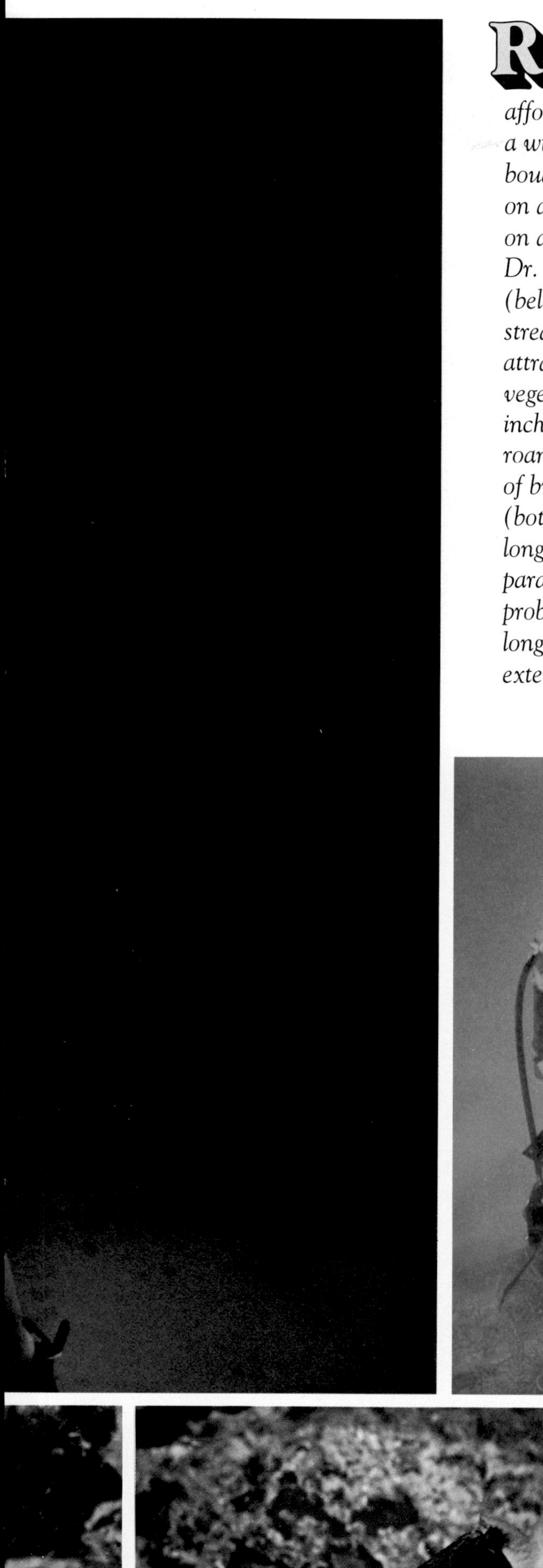

Room with a view, Tektite's hemispheric window (left) affords Peggy Lucas, habitat engineer, a wide-angle look at Sylvia Earle's bouquet of algae. Comparing life on a natural bed of sea grass with that on a patch of artificial greens, biologist Dr. Renate True and Dr. Earle (below) discover that the plastic streamers, which algae soon covered, attract animals as readily as real vegetation. Zebra stripes band a six-inch-long jackknife (bottom, left) roaming near Tektite. Close to a head of brain coral, a tiny sharknose goby (bottom, right) cleans a coney—a foot-long grouper—by picking off external parasites. Habitats help scientists to probe deeper and remain underwater longer—allowing them to make extended studies of the marine world.

ALVIN

Windows Beneath the Sea

"Now as before, we were dropping down, down, down, at 200 feet per minute. . . . Black water rushed upward past us. . . . At 20,000 feet, we were at the maximum depth of the normal Pacific sea floor. We were dropping into the open maw of the Mariana Trench, leaving the abyssal zone of the ocean and entering the hadal regions."

So wrote Swiss oceanographic engineer Jacques Piccard after his historic descent in late January 1960 with Lt. Don Walsh, USN, in the bathyscaph *Trieste.* Finally, after an eerie, incredible voyage into the unknown, *Trieste* settled 35,800 feet down in the deepest known place in the sea. Piccard continued:

"Indifferent to the nearly 200,000 tons of pressure clamped on her metal sphere, the *Trieste* balanced herself delicately on the few pounds of guide rope that lay on the bottom, making token claim, in the name of science and humanity, to the ultimate depths in all our oceans—the Challenger Deep."

It had taken nearly four centuries for mankind to plunge some seven miles, from the time of the first detailed, practical description of a submarine, penned by William Bourne in 1578, to the remarkable achievement by Piccard and Walsh.

The advantages of travel in submarines were clear to some as early as the mid-1600s, when Englishman John Wilkins wrote:

"1. 'Tis *private;* a man may thus goe . . . without being discovered. . . .
2. 'Tis *safe;* from the uncertainty of *Tides,* and the violence of *Tempests.* . . .
3. It may be of very great advantage against . . . enemies, who by this means may be undermined in the water and blown up.
4. It may be of a special use for the relief of any place that is besieged by water, to convay unto them invisible supplies. . . .
5. It may be of unspeakable benefit for submarine experiments and discoveries. . . ."

Wilkins knew of Cornelius van Drebbel, who designed what is thought to be one of the first workable undersea vessels. According to English scientist Robert Boyle, van Drebbel's submarine used "a chemical liquor, which he accounted the chief secret of his submarine navigation. For when, from time to time, he perceived that the finer and purer part of the air was consumed, or over-clogged by the respiration and steames of those that went in his ship, he would, by unstopping a vessel full of this liquor, speedily restore to the troubled air such a proportion of vital parts, as would make it again, for a good while, fit for respiration." This seems to be the first submarine with a renewable supply of air—a vital key to submersible success.

". . . the only other place
comparable to these marvelous
nether regions, must surely be
naked space itself, out far beyond
atmosphere, between the stars . . .
where the blackness of space,
the shining planets, comets, suns,
and stars must really be closely
akin to the world of life as
it appears to the eyes of an awed
human being, in the open ocean,
one half mile down."

William Beebe,
Half Mile Down

Laden with instruments and remote-control samplers, Alvin *explores the ocean floor during a 1979 expedition to the Galapagos Rift off the coast of South America. One of the most widely used research submersibles ever built,* Alvin *has dived a thousand times since her commissioning in 1964. Such vessels have made unparalleled contributions to the study of earth's "inner space."*

NATIONAL GEOGRAPHIC PHOTOGRAPHER EMORY KRISTOF AND ALVIN M. CHANDLER, NATIONAL GEOGRAPHIC STAFF, BY REMOTE-CONTROL CAMERA

In 1775 David Bushnell, an American, designed the small wooden submarine *Turtle* for use against the British in the Revolutionary War. Most of the submarines that followed in the next century, those of Fulton, Bauer, and Holland, were developed primarily for combat. But another motive—curiosity—caused naturalist William Beebe to

DAVID KNUDSEN, COURTESY DR. WILLIAM BEEBE

First modern submergence chamber built for direct observation of the ocean depths, the bathysphere of pioneering deep-sea researcher Dr. William Beebe hangs from a boom on its support barge off Bermuda in 1934. The steel sphere, only 4 feet 9 inches in diameter, snugly held two men, oxygen tanks, a telephone, and a searchlight.

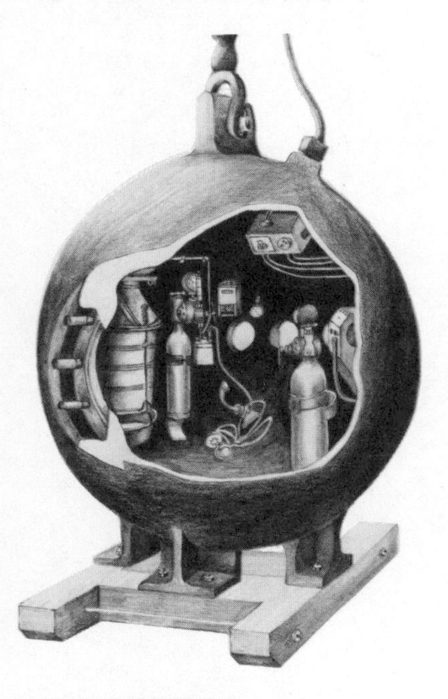

MARIANNE RIGLER KOSZORUS,
FROM DRAWING BY CHARLES E. RIDDIFORD

look for a way to go deeper in the sea than anyone had gone before. Beebe recalled the birth of his bathysphere in *Half Mile Down:* "During 1927 and 1928 I considered various plans for deep-sea cylinders . . . since there is nothing like a ball for even distribution of pressure, the idea of a perfectly round chamber took form and grew. By 1929 Mr. Otis Barton had developed . . . a steel sphere."

Lowered and raised by a steel cable fixed to a winch on a barge, the submersible contained enough oxygen to sustain two people for eight hours as well as a tray of carbon dioxide absorber. Beebe and Barton made a series of successful dives in the bathysphere near Bermuda in June 1930, reporting back to the surface via telephone:

" 'We have just splashed below the surface.'
'We are now at our deepest helmet dive.' 60 feet
'The *Lusitania* is resting at this level.' 285 feet
'This is the greatest depth reached in a regulation suit by Navy divers.' 306 feet
'We are passing the deepest submarine record.' 383 feet . . .
'A diver in an armored suit descended this far into a Bavarian lake—the deepest point which a live human has ever reached.' 525 feet
'Only dead men have sunk below this.' 600 feet
'We are still alive and one quarter of a mile down.' 1426 feet"

At that depth Beebe said that his diving partner, Barton, was "droning out something, and when it was repeated I found that he had casually informed me that on every square inch of glass on my window there was a pressure of slightly more than six hundred and fifty pounds. . . . After this I breathed rather more gently in front of my window and wiped the glass with a softer touch, having in mind the nine tons of pressure on its outer surface!"

Barton calculated the total pressure on the small sphere. The result was 3,366.2 tons, more than six and a half million pounds.

In 1934 Beebe and Barton had reached a depth greater than half a mile in the bathysphere, observing along the way black eels, golden-tailed sea-dragons, larval fish. Beebe also reported, "Amid nameless sparks, unexplained luminous explosions, abortive glimpses of strange organisms . . . a definite new fish or other creature. . . ." Beebe's descriptions, given over the bathysphere's telephone to the surface, provided the first eyewitness accounts of life in the deep sea.

The bathysphere realized Wilkins' prediction of being "of unspeakable benefit for submarine experiments and discoveries," but its range was dependent on the length of a single seven-eighths-inch cable by which it was reeled from its surface support ship into and out of the sea.

The next attempts to enter the abyss took place in vehicles called bathyscaphs, designed by Auguste Piccard beginning in 1937. Operating like underwater balloons, bathyscaphs went ever deeper, and one, *Trieste,* eventually plunged into the Mariana Trench, setting a record that remains unbroken.

During the 1950s Capt. Jacques-Yves Cousteau became interested in combining the ease of free diving with the advantages of a pressure-resistant shell for excursions below 300 feet. "But a regular submarine," Cousteau said, "or a deep-diving bathyscaphe, would be too big and clumsy for intimate reef exploration. We needed a radically new submarine, something small, agile."

By 1960 Cousteau had pioneered the development of a remarkable diving saucer without rudder or tail planes that could jet around reefs and along cliff faces to 1,000 feet. Dr. Harold Edgerton, designer of much of the sophisticated camera equipment used in the ocean depths, logged impressions of his maiden voyage in Cousteau's saucer: "Like an airplane we descended to where the reef dropped off into deeper water. . . . We were getting good, clean oxygen. Being in the saucer was no different from being in an automobile, except that we had more room and lolled comfortably on our foam mattresses like Romans at a banquet. . . . Falco [the pilot] spotted a squadron of squid, swimming against the bottom in perfect formation. . . . A host of fishes of many colors circled us. . . ."

Fast, deep-diving submarines were being built for use by naval forces in several countries. The most sophisticated were nuclear-powered ships that could operate submerged for months. In March 1959 one of these, the U. S. nuclear submarine *Skate,* surfaced at the North Pole, breaking through the ice 50 years after Comdr. Robert E. Peary laboriously attained that point overland by dogsled. On an earlier cruise, as *Skate* was approaching the Arctic Ocean, men aboard had heard the following radio broadcast: "The entire civilized world thrilled today to the announcement that the American atomic submarine *Nautilus* has crossed from the Pacific to the Atlantic under the Arctic ice pack, sailing under the North Pole en route. . . ."

Skate's captain, Comdr. James F. Calvert, witnessed the reaction to the news: "A gasp of incredulity went up from everyone in range of the radio. I was the only one on board who had been authorized to know in advance of the ultra-top-secret voyage of the *Nautilus.* Our primary missions were quite different—*Skate*'s to learn surfacing in the ice, *Nautilus*'s to pass from Pacific to Atlantic beneath it."

Military submarines are generally built to move undetected as self-contained weapons carriers made of steel. The 447 1/2-foot-long nuclear submarine *Triton* succeeded admirably when she made a 30,752-mile round-the-world underwater cruise in 61 days in 1960.

But three years later tragedy struck the nuclear submarine force. *Thresher,* apparently out of control, sank inexplicably to a depth of

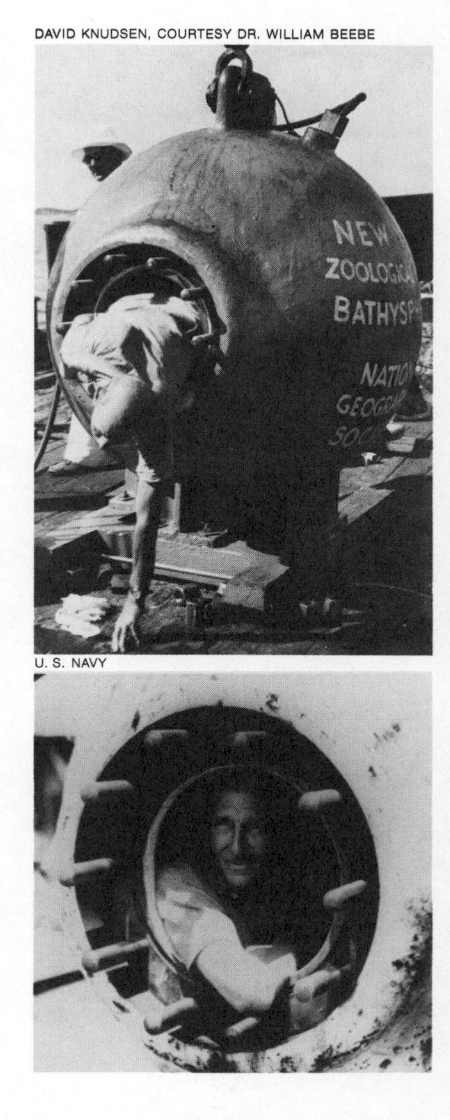

DAVID KNUDSEN, COURTESY DR. WILLIAM BEEBE

U. S. NAVY

Naturalist Beebe peers through the 14-inch opening of his bathysphere (above) before carefully squeezing past the bolts that seal the 400-pound door (top) and emerging into the sunlight. In 1934, during an expedition to Bermuda sponsored by the New York Zoological Society and the National Geographic Society, Beebe and designer Otis Barton descended 3,028 feet, setting a record that remained unbroken for 15 years.

8,400 feet 260 miles off the coast of New England. There was no way to conduct a rescue or recovery. Only one U. S. craft, the bathyscaph *Trieste,* could even descend to the site to discover what had become of the ill-fated *Thresher.* And just two other vessels, the French bathyscaphs *F.N.R.S. 3* and *Archimède,* had ever been that deep. *Trieste* found and photographed the sunken submarine, but salvage, though considered, was found to be impossible.

COURTESY DR. JACQUES PICCARD

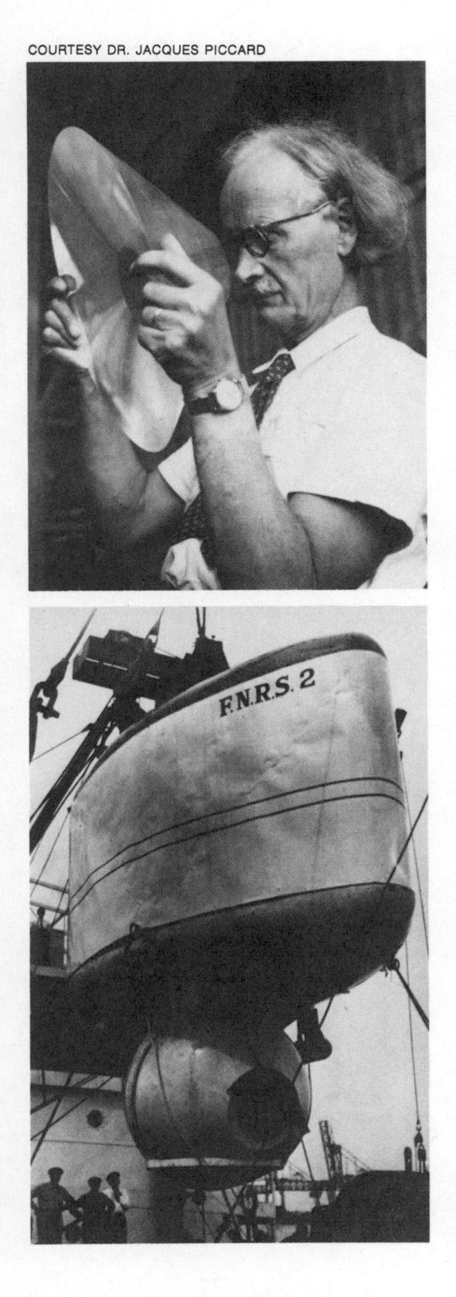

James H. Wakelin, Jr., then Assistant Secretary of the Navy for Research and Development, summed up the paradox of the times in a NATIONAL GEOGRAPHIC account in June 1964: "In the years of uneasy peace following World War II, U. S. naval technology gave the fleet a new kind of submarine—fast, deep diving, nuclear powered. Within less than a decade our Navy was projected into a new and virtually unknown environment. Submarines had been limited to the upper few hundred feet of the ocean; now they ranged to considerable depths, into a frontier where our oceanographic knowledge had not kept pace with the swift development of shipbuilding technology."

A host of small submersibles were quickly developed to close that gap. Among the first such vessels built in the United States were those designed by entrepreneur John Perry. In 1962 his *Cubmarine,* a two-man craft capable of descents to 300 feet, was launched on a career that included use by the U. S. Army to recover spent missiles from a testing range in the Pacific. A 1971 Perry creation, the *PC-8B,* introduced a distinctive nose that provided unusually good viewing capability. The feature has been incorporated into larger subs that can routinely dive to 1,200 feet for use in offshore oil-well maintenance.

Among the smallest research submersibles was the 16-foot *Asherah,* built for the University of Pennsylvania specifically for archaeological studies. She proved her worth in 1964 at a Roman wreck off the coast of Turkey in 140 feet of water; in less than an hour she took dozens of stereophotographs—a job that would have taken diving teams weeks to accomplish.

One of the largest and most extensively used research submarines is Reynolds Metals' *Aluminaut,* a 51-foot deep-diving vessel that operates with a complement of seven: three crew members and up to four observers. Her career has included scientific, military, and industrial operations since she was launched in 1964. But the best remembered achievement was probably one poignant mission in 1969 beneath the cold, dark waters of the North Atlantic. There the Woods Hole Oceanographic Institution submersible *Alvin* rested on the bottom for almost a year following an accident that resulted in the flooding and sinking of the 22-foot-long craft. *Aluminaut* descended nearly a mile to a site pinpointed by a sonar "fish" towed by a surface ship. Then, using *Aluminaut*'s mechanical arms and tools, the pilot artfully maneuvered a large toggle bar into *Alvin*'s hatch and wedged it firmly into place across the opening so the submersible could be recovered.

Another relatively large research submarine, the 16-by-40-foot *Deep Quest,* was built by Lockheed to accommodate a payload of 7,000 pounds of equipment, along with two pilots and two observers. Used for geological research, *Deep Quest* also completed several salvage jobs, including the recovery in 1969 of the "black box" flight recorders from two airliners that had crashed near Los Angeles.

Most small submersibles are meant for use on brief missions of 3 to 6 hours. The *Ben Franklin,* a 49-foot steel-hulled vessel with 29 acrylic view ports, is a notable exception. Designed for free drifting at depths to 2,000 feet, the submersible undertook a historic mission of oceanographic exploration in 1969; six men, including expedition leader Jacques Piccard, were aboard for the 1,500-mile, 30-day journey deep within the Gulf Stream.

Initially research submersibles were greeted with skepticism by many scientists, who quite reasonably asked, "What can be accomplished by taking people underwater that cannot be done as effectively by instruments lowered from the surface?" The answer came with use. After marine geologist Dr. Conrad Neumann made his first dive in *Aluminaut,* from which he was able to observe large areas of the seafloor at once—not just in photograph-size pieces—he said: "My opinion of deep-diving vehicles has reversed from one of negative skepticism to enthusiastic optimism. I feel the manned submersibles are the best thing to come along in oceanography since galvanized wire and electrician's tape."

Conqueror of the greatest ocean depths, the bathyscaph Trieste *(above) works like a balloon filled with air. Its striped float contains aviation gasoline that provides buoyancy underwater; below hangs the cabin, a steel sphere 7 feet 2 inches across.* Trieste's *designer, Swiss scientist Auguste Piccard (opposite), holds an acrylic plastic window he developed for his first submersible,* Trieste's *predecessor,* F.N.R.S. 2.

THOMAS J. ABERCROMBIE, NATIONAL GEOGRAPHIC STAFF

In all, more than 100 shallow and deep-sea research vessels have been constructed in 14 countries since 1960, some for industrial use in connection with recovery of oil and other minerals from the sea and some specifically for scientific exploration. All have distinctive shapes and capabilities. About these qualities underwater writer Robert Burgess wryly observed, "Most laymen can recognize them on sight because '. . . most of them look like something from outer space, and they cost a fortune to build.' "

In August 1975 I was perched happily in one of these inner-space craft. Seated as an observer in the acrylic sphere of the *Johnson-Sea-Link I,* I waved to the submersible's designer, Edwin A. Link, who was piloting *JSL-I*'s twin, the *JSL-II.* In response Link plucked a small bouquet of sponges from the seafloor with his submersible's manipulator and ceremoniously placed it in the collecting basket of *JSL-I.* He then managed a remarkably eloquent mechanical wave before moving on to an area where our serious search-and-recovery exercise of the day was to take place.

Accompanying us on the ocean bottom was a Cabled Observation and Rescue Device, an unmanned vehicle that is operated from the surface. Equipped with a television "eye," a mechanical "hand," and a powerful rotating saw capable of cutting through steel cable, the

CORD is one of several machines conceived by Link and built for exploration and submarine rescue.

After the *Thresher* tragedy, Link had joined a group that helped the U. S. Navy analyze and strengthen the nation's undersea capabilities and recommended ways to develop Navy rescue systems.

Link proceeded on his own to design other submersibles, and in 1970 he set up the Harbor Branch Foundation, Inc., a marine research facility in Link Port, Florida. By the end of the year a new submersible, dubbed *Johnson-Sea-Link I,* was ready for launching. The grave urgency of mastering rescue technology was heightened after Link's younger son, Clayton, and a companion, Albert Stover, died when the *JSL-I* was trapped in the wreckage of a sunken ship at 360 feet.

I HAD FIRST MET Ed Link and his wife, Marion, several years before our sponge-gathering excursion, when I was a participant in the 1968 Smithsonian-Link Man-in-Sea Project. For several weeks I enjoyed their hospitality aboard the 100-foot-long *Sea Diver,* and I looked forward to using the remarkable new Perry-Link *Deep Diver* submersible that sat in a cradle on the ship's deck. This vessel carried a pilot and an observer in a forward compartment and also could hold two divers in an aft lock-out section. In that rear compartment air could be compressed to equal outside pressure down to 1,250 feet. The concept was brilliant.

I remembered Assistant Editor Kenneth MacLeish's impressions of the sub in a NATIONAL GEOGRAPHIC account when I found myself sitting in exactly the same spot in *Deep Diver* that MacLeish had occupied. Tongue-in-cheek, he had written, "I pondered the question of what in the deep blue sea I, an aging writer and downright elderly diver, was doing in the hind end of a bottomward-bound diving machine . . . about to be spawned like a fishlet into a fish's world."

Completing her maiden voyage, the nuclear submarine Scorpion *returns to her berth in Groton, Connecticut, in 1960. Vice Adm. Hyman G. Rickover, the father of the atomic submarine fleet, stands at left with the ship's Division Commander, James F. Calvert. Tragedy struck the* Scorpion *eight years later when she failed to return from an undersea mission. After a five-month search, a remote-control camera on an undersea sled photographed the hull in more than 10,000 feet of water near the Azores. Below, a mooring line protrudes from its stowage locker in* Scorpion's *after section. In early 1969 the U. S. Navy sent* Trieste II *to the wreck site, where the bathyscaph surveyed and photographed the nuclear sub. Such missions underscored the military usefulness of research with deep-diving submersibles.*

U. S. NAVY

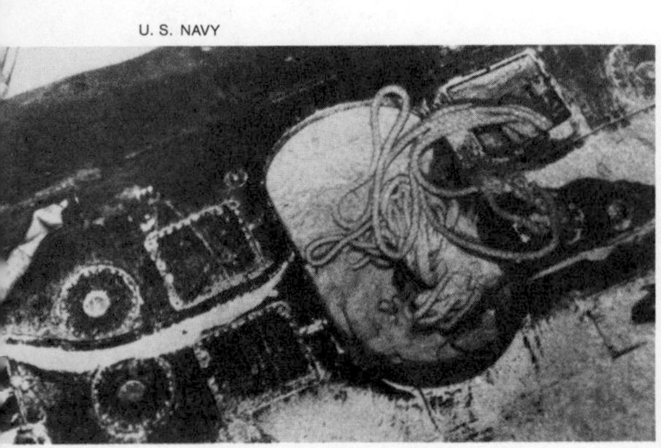

Like an undersea fishbowl, the acrylic plastic sphere of the Naval Experimental Manned Observatory provides two shirt-sleeved observers wide-angle views of depths as great as 600 feet. At its destination, NEMO releases an anchor, then winches itself up and down like an undersea elevator. The bottom sections hold batteries, lights, hydraulic equipment, and electric motors.

BLACK STAR

Ed Link explained *Deep Diver* to MacLeish as "more than a sub. . . . It's a system. It includes a special hydraulic crane that can launch and retrieve her even in rough weather. . . . We park her on the bottom and build up the gas pressure in the divers' compartment until it equals the pressure outside; the hatch drops open and out they go.

"When they get back in, they simply close their hatch to lock the pressure in with them, and come on up. We hoist her aboard and sail away. . . ." Decompression occurs on board, inside the sub.

On his dive, MacLeish, along with chief engineer and veteran diver Dennison Breese, locked out at 420 feet. Later, the two decompressed for three hours in the sub's aft chamber.

NOW I WAS CONCENTRATING on *my* first lock-out trip in a submersible. After only a few minutes underwater, I earnestly vowed that it would not be my last encounter with underwater taxi service. We descended smoothly along a gradual slope, and through the small port of the diver's compartment I watched the transition into deep indigo. Then I felt a mild bump as we gently settled at the lock-out site. Denny Breese, my partner, communicated with the pilot in the forward compartment: "We're ready when you are."

He turned a valve to increase the inside pressure until it equaled that of the surrounding sea. Compressed air hissed and swirled around us. My ears popped as I swallowed and cocked my jaw to equalize the changing pressure—just as I do when flying or making a long, fast descent in an elevator.

Denny loosened the bars on the hatch at our feet; it groaned softly, then swung down, offering us a liquid door to the sea beyond. He smiled at my incredulous expression. We were nearly 100 feet underwater, yet I could stand on the seafloor with water to my waist and my head and shoulders in the warm, dry atmosphere of *Deep Diver.*

Denny grinned and suggested, "Don't keep the fish waiting."

I glided out over a soft, sandy bottom and into an amphitheater of blue. At first I kept raising my head from the plants and fish I had come to observe, glancing back at the yellow submersible that waited for me. *Deep Diver* seemed to have its own personality; the ports and diving planes gave it a benevolent, bemused expression.

But the plants soon won my undivided attention, and an hour and a half passed too quickly. It was longer than I had ever been able to stay at that depth, and never before had there been a decompression chamber standing by on the seafloor awaiting my return! On the way back to *Sea Diver* Denny began to reduce pressure, a procedure that continued for some time after the special crane lifted the sub out of the sea and gently placed her in the on-board cradle.

As I sat within the steel cocoon, I mused about what the child I was then carrying might think when I could describe the experience to him or her some years later. I was five months pregnant, but doctors I had consulted foresaw no difficulties, and there were none. Today my youngest daughter beguilingly explains her passion for the ocean: "I can't help it. I was diving from a submarine before I was born."

During that Man-in-Sea program, Ed Link described another idea to me, one that would make it possible for the first time to conduct open-water lock-out dives along steep cliffs, where there is no parking space for a traditional lock-out submersible. Link reasoned that if the craft could be held steady enough, it could function like a diving bell, open at the bottom. Divers would swim out to work, then return to the sub. A special *(Continued on page 175)*

NATIONAL GEOGRAPHIC PHOTOGRAPHER EMORY KRISTOF

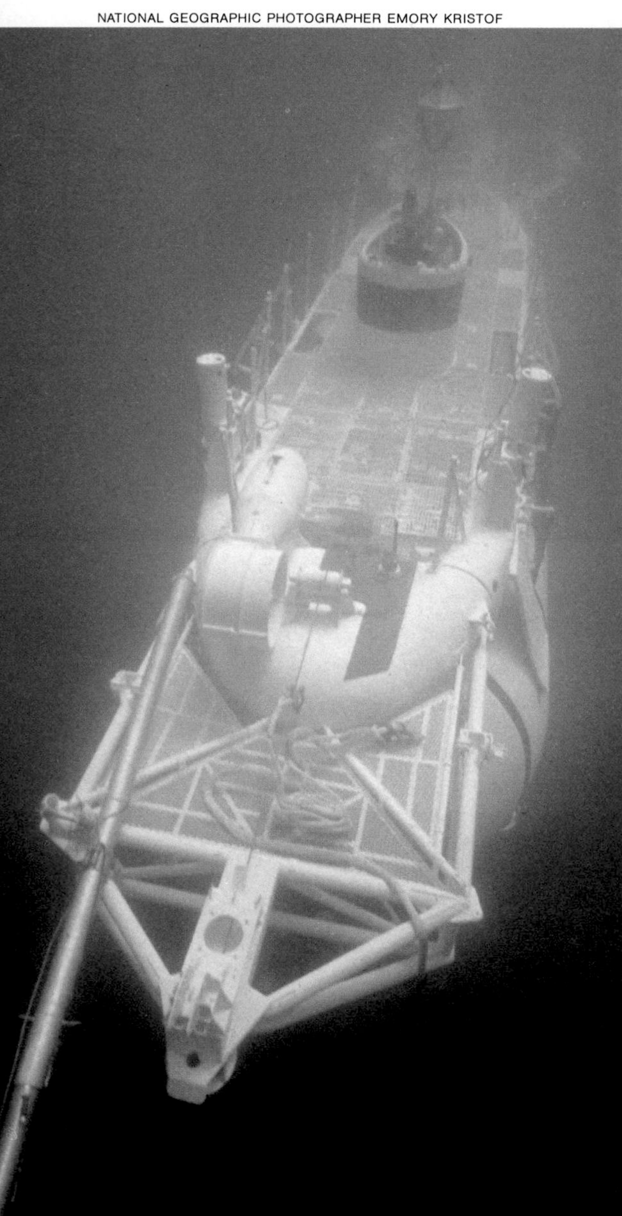

Floodlights pick out Trieste II *as she approaches her berth in the support ship* Point Loma *after a 1977 dive into the Cayman Trough. Dropping into the gloom (left), the bathyscaph descends more than 20,000 feet into the Caribbean to the bottommost volcanoes in the world. In 1963 the*

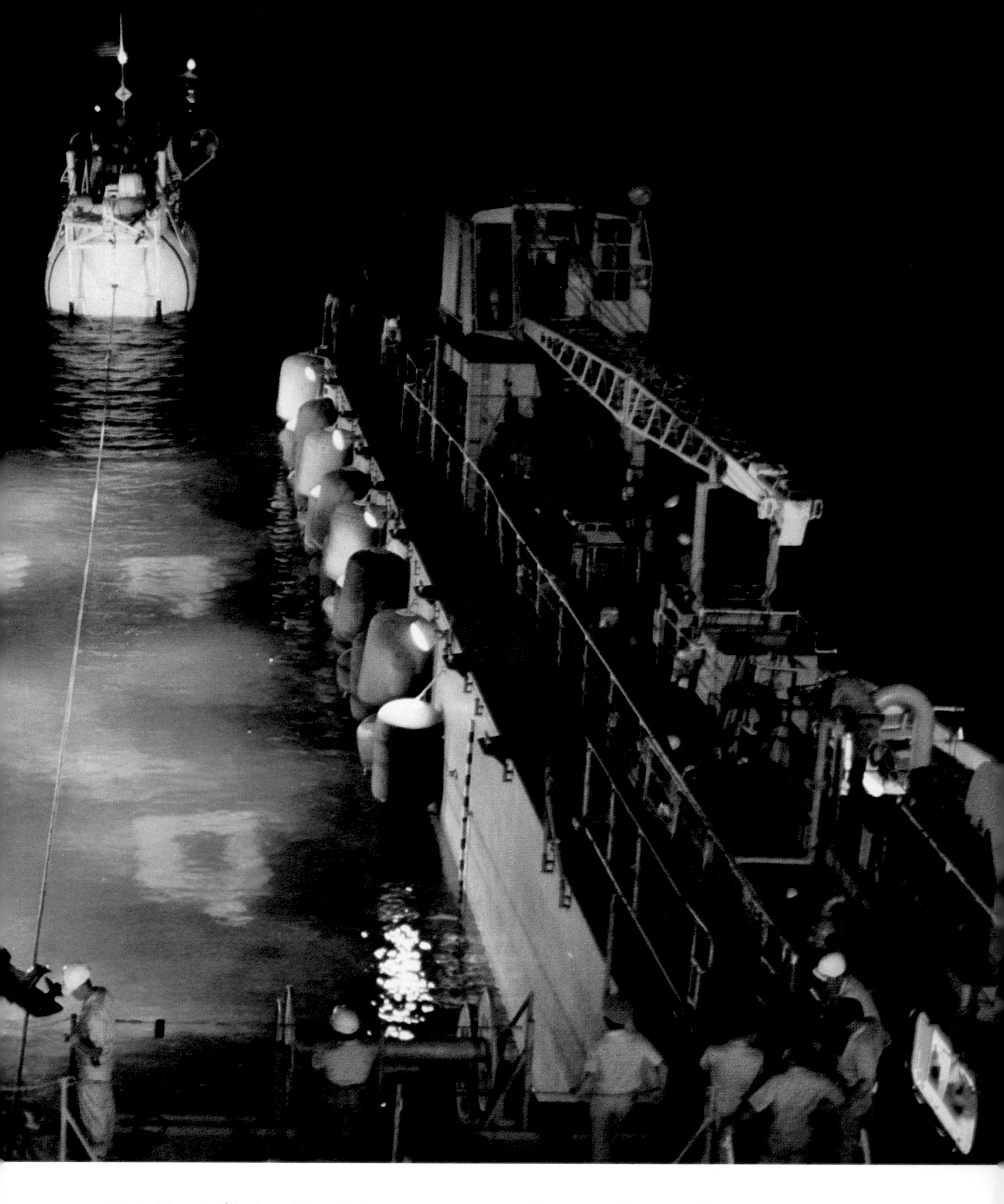

U. S. Navy had built and launched Trieste II, *an improved version of the Piccard bathyscaph.*
PRECEDING PAGES: *Buffeted by waves that tore loose a towline,* Trieste *undergoes repairs en route to the Mariana Trench. On January 23, 1960, Jacques Piccard (at right) and Lt. Don Walsh,* USN, *took the craft 35,800 feet into the trench, the deepest known spot in the sea. As they reached the ocean floor a fish swam away, proof that living creatures exist in the depths.*

THOMAS J. ABERCROMBIE, NATIONAL GEOGRAPHIC STAFF (PRECEDING PAGES)

NAVY

anchoring system would be needed. To illustrate the point, Link sketched a palm tree with a chain wrapped around the base. The chain dangled over a steep drop-off into the sea; *Deep Diver* was pictured suspended at the end.

Several years later I again looked through a liquid blue doorway in the aft end of a lock-out submersible. But this time I peered out of Link's latest creation, the *Johnson-Sea-Link II.* Below, there was no reassuring white sandy bottom—only 1,000 feet of indigo infinity.

Link's vision had come to pass. The submersible was suspended from an anchoring cable firmly implanted in the reef above, and I was free to glide out of the sub for nearly an hour of exploration at 250 feet along a steep cliff face.

My diving partner, Steve Nelson, helped me strap on a Kirby-Morgan helmet that fed air past my face in a soft, steady stream. I adjusted my flippers and slipped quietly into the blue world beyond on a brief, but welcome, passport. A few feet away I touched one finger to the cliff to support myself while I looked back at a midwater apparition. Two men, like blue Wizards of Oz, seemed to float in their acrylic sphere, illuminated by several glowing lights mounted on the sub.

"Sylvia, you are now three minutes into your dive."

I saw chief pilot Jeff Prentice wave from the acrylic globe and replied through my communication system, "Roger, three minutes."

I imagined myself a hawk, soaring effortlessly in blue space . . . but remembered I was a botanist after all, as I spotted a diminutive forest of plants that I recognized as something new. Each elfin grove resembled a group of miniature green parasols flourishing on a narrow ledge in an atmosphere of sapphire light.

I had just begun to touch and think and count, noting the small animals among the plants, when Jeff reminded me that it was time to reenter the submersible and start decompression.

Instead of taking us to the surface, however, the *Johnson-Sea-Link* traveled a few hundred feet over the reef and deposited Steve and me on the doorstep of another apparition—the sturdy yellow form of Hydro-Lab, our home-on-the-reef for nearly a week. Inside, our companions, zoologist Dr. Thomas Hopkins and German scientist Gerd Schriever, were writing up their observations from a recent 200-foot-deep, hour-long swim over the edge of the cliff.

We were taking part in the Scientific Cooperative Operational Research Expedition, jointly sponsored by the Harbor Branch Foundation, the Perry Foundation, and the National Oceanic and Atmospheric Administration. Project SCORE was to demonstrate the advantages of saturation diving coupled with a diver-transport system: extended time at extended depth and the ability to decompress in the submersible on the way home. From the surface, an hour-long dive to 200 feet would require 200 minutes of decompression. Since we were starting from and returning to Hydro-Lab at 60 feet, however, our decompression time was only 19 minutes.

NOT ALL SUBMERSIBLES now constructed include a diver lock-out capability. For some the focus is not on releasing divers, but on developing other parts of the system—manipulators, remote cameras, elaborate sensors.

One submersible that boasts such features is *Alvin,* operated since her commissioning in 1964 by the Woods Hole Oceanographic Institution. I first met *Alvin*—and Allyn Vine, one of her originators—in a shed at Woods Hole in 1974. "She looks like a disassembled beetle," I

Testing the Mark I Deep Diving System, U. S. Navy divers release the mechanism that swung the personnel transfer capsule off the support ship. Developed in 1970, the system enables two of the three divers inside the sphere—lowered to the depths by an additional cable—to swim out on a salvage or rescue mission. Hoisted back aboard ship, the capsule connects with a deck decompression chamber. During trials of the system off Panama City, Florida, in June 1975, Navy divers made the deepest plunge in the open sea until then: 1,148 feet.

BERNIE CAMPOLI

observed, noting winglike portions here, "antennae" there, and a glistening large ball in the midst of it all.

"The most important part remains intact," Vine pointed out, touching *Alvin*'s shining new titanium sphere with obvious affection. *Alvin* was undergoing routine servicing, having recently returned from

COMDR. JAMES F. CALVERT, U. S. NAVY

Wan Arctic sun greets U. S. nuclear submarine Skate, *only 300 miles from the North Pole on March 22, 1959. During this historic voyage* Skate *surfaced through 10 "skylights"—thin spots in the 12-foot-thick ice cover. One breakthrough took place precisely at the Pole.*

participation in a historic series of dives during Project FAMOUS, the French-American Mid-Ocean Undersea Study. The project provided an incentive for the first full-scale use of manned submersibles for deep-sea research and contributed invaluable evidence to support the theory of plate tectonics.

First to dive in the course of the project was the French Navy's bathyscaph *Archimède,* capable of withstanding the pressure of the ocean's greatest depths. A year later *Alvin* and the French diving saucer *Cyana* joined forces with *Archimède* for a total of 51 dives to explore the earth's changing crust.

These three submersibles complemented the international surface fleet that, over a two-year span, probed a 30-mile section of the 12,000-mile-long Mid-Atlantic Ridge. That ridge, in turn, joins the Mid-Oceanic Ridge system, which winds its way along the bottom of the seas around the globe—40,000 miles in all. Dr. J. R. Heirtzler, U. S. Chief Scientist of Project FAMOUS, describes the system as "the largest mountain range on this planet—a system greater than the Rockies, the Andes, and the Himalayas combined." The Mid-Atlantic Ridge is the most rugged of the entire system. Along its crest is a central rift where new ocean floor is being poured forth—and carried away with drifting crustal plates.

The continuing quest for information on plate tectonics lured *Alvin* scientists to the Pacific for the first time in 1977. They got more than they bargained for in the Galapagos Rift. Near warm-water vents at depths of 8,000 to 9,000 feet, observers were astonished to see bizarre organisms—giant clams, pale crabs, mysterious white threadlike worms and dandelion-shaped creatures that no one could identify, as well as long tube worms that did not fit any textbook description.

An ambitious expedition was immediately proposed; it began two years later. The 1979 project, funded by the National Science Foundation and the Office of Naval Research, would combine geological and biological research. Teams of research scientists led by geologist Dr. Robert D. Ballard, biologist Dr. J. Frederick Grassle, and geo-

GEOGRAPHY AND MAP DIVISION, LIBRARY OF CONGRESS

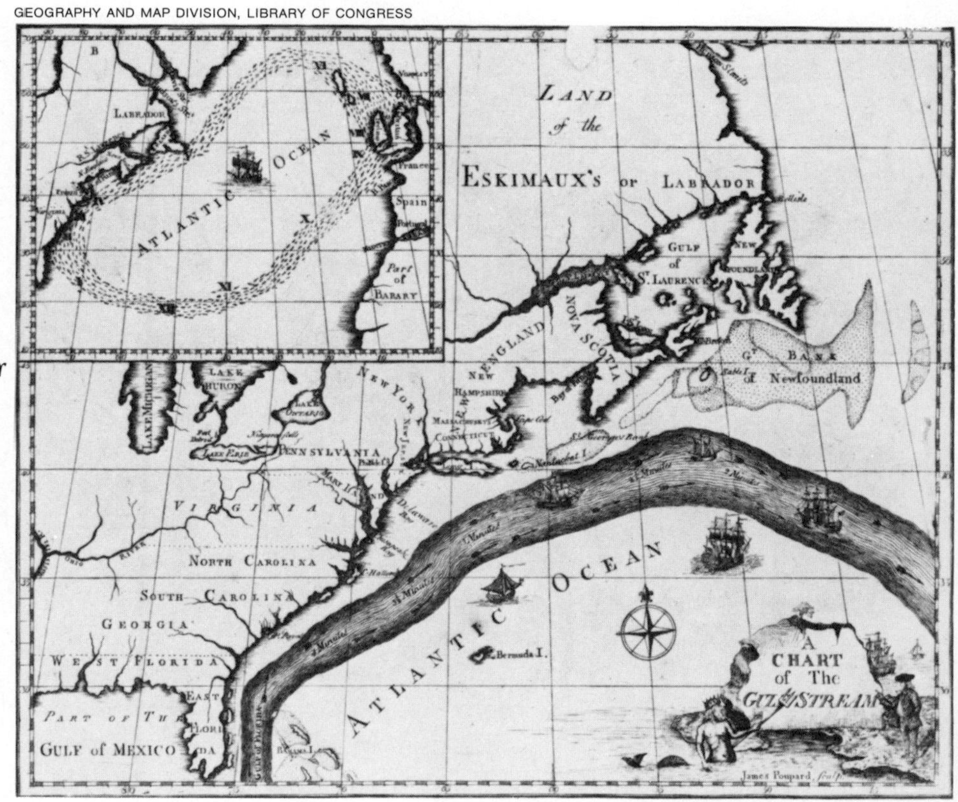

Neptune encounters a landsman on an early chart of the Gulf Stream. Deputy Postmaster General Benjamin Franklin commissioned the map after he received complaints that it took longer to receive a reply from England than it did to send a letter there. A Nantucket whaling captain provided the explanation: The current delayed ships sailing westward by up to 70 miles a day. Two centuries later scientists aboard the submersible Ben Franklin *carried a copy of this map with them as they explored the Gulf Stream.*

chemist Dr. John Edmond would use *Alvin* and special new photographic equipment to observe the strange ecosystem. In a series of dives to be sponsored by the National Geographic Society, Al Giddings was asked to co-direct a film of the creatures encountered in the warm-water vents, using an innovative television camera.

"What do you think?" Al asked me. "Should I go?"

I knew he was heavily committed to other projects, especially the conversion of the research vessel *Eagle* for an extensive scientific and photographic program. Al's departure for three months would bring certain aspects of work on *Eagle,* already delayed, to a standstill. But there could be only one response to Al's question. Certain of what he would decide anyway, I gently asked him, "Would you refuse a ticket to the moon?"

No hero's welcome greeted Al on his return from his visits to the ocean floor. The feelings he expressed to me upon reentry to surface life in San Francisco were reminiscent of Lt. Comdr. Georges Houot's comments after his 59th trip to the black abyss in the French bathyscaph *F.N.R.S. 3:* ". . . I am driving home for dinner. . . . My glance roams over my neighbors sitting at their wheels, but my eyes do not see. My thoughts are elsewhere, in another world, strange and fascinating, which only a few privileged persons have seen. . . .

"Only a few hours ago I was off the coast, 7,500 feet under the sea. Yet no one looks at me; I bear no visible trace."

Returned from *his* voyage, Al ceremoniously placed a small, glassy, black chunk in my hand.

"For you," he said, smiling. "A 'moon rock' from the depths of the Galapagos."

The glistening fragment from the seafloor brought to Al's mind visions still fresh of the wonders he had seen in the ocean depths.

"The crabs were unbelievable!" he said. "We watched them tiptoe up to the giant tube worms and snip off bites before the worms could retreat into their shelters! Most beautiful was the Rose Garden. There were hundreds of red-plumed tube worms, some nearly as big around as your wrist."

Insider's view: Water splashes against the acrylic sphere of the Johnson-Sea-Link I *as her pilot takes the craft down. Edwin A. Link, pioneer inventor and researcher in aeronautics and ocean engineering, launched the vessel in 1971.*

LUIS MARDEN

The animals Al described belong to a complex ecosystem, all of them heretofore unknown and the most conspicuous—giant tube dwellers—so different that they are being classed as a separate subphylum of animals. A significant discovery is that large amounts of hemoglobin—the oxygen-carrying substance in human blood—courses through the bodies of the Galapagos tube worms and accounts for the red color.

The most exciting biological discovery concerning the newfound ecosystem is how it seems to sustain itself. On land, plants derive energy from the sun through photosynthesis. But here, close to the vents, the basic food chain begins instead with sulfur bacteria that obtain their energy from a chemical reaction. Such systems, moreover, may not be so rare in the deep sea, if—as can now be assumed—more vents exist. Later dives in *Alvin* in 1979 near the tip of Baja California resulted in the discovery of another system, including many groups of organisms similar to those in the Galapagos Rift.

The Baja dives caused even greater excitement when geologists observed chimneys of rock spewing forth black clouds of minerals: iron, copper, zinc, and silver sulfides. "It may be that minerals we find on dry land today were formed by a process such as we witnessed," Robert Ballard commented. "The deep-sea discoveries made during the coming decade should help solve more of the mysteries of earth's origin—and may provide valuable new sources of minerals as well." No net or dredge could catch what geologists in *Alvin* witnessed at Baja and brought to the surface on film and in their minds.

IN EARLY 1979, in Switzerland, I sat drinking hot tea and eating croissants across the table from a man who holds numerous depth and duration records, Jacques Piccard. Outside, a gray sky and drizzling rain softened our view of Lake Geneva and Piccard's newest submersible, the *F. A. Forel.* We had just returned from a brief excursion that gave me a chance to admire the exceptional viewing capability of the sub and the magic whereby Piccard comfortably bends his long, agile body into an amazingly compact space.

"Where from here?" I asked, curious about his next submersible design, already under way. "Not deeper?" Submersibles in 1980 obviously need go no deeper than *Trieste* went 20 years ago. "Not longer?" There may be little point in staying submerged longer than the many months already possible in nuclear submarines.

Piccard smiled, brown eyes lighting as he answered, "No. Better!" He went on to describe his vision of "a swift submarine; one that will travel long distances on its own, one that will be able to go deep, to go where eels spawn in midocean, under the Sargasso Sea; one that can follow whales."

Piccard's comments reflected an attitude he expressed after his deepest *Trieste* dive: "The ocean imposes modesty on those who try her but basically the deep sea is friendly. . . ." He adds, "That [man] plans to re-enter the sea with tools and vehicles of his own devising is a measure of his incredible success. That man is bent on ultimate adventure at the basement of the earth, there is no doubt at all."

N.G.S. PHOTOGRAPHER BATES LITTLEHALES (ABOVE); SYLVIA A. EARLE (BELOW)

Helmeted swimmer photographs Deep Diver *in shallow Bahamian waters during a practice dive in 1967; scuba divers observe the plunge. Divers could leave and reenter the submersible underwater through a lock-out chamber. Within days* Deep Diver *carried two men 420 feet down for the deepest lock-out dive made until then.* Deep Diver's *designer, Edwin A. Link, included the chamber in the* Johnson-Sea-Link I, *his next submersible. Its transparent hull offers pilot and observer excellent visibility. The* Makakai *(below), launched in 1971 by the U. S. Naval Ocean Systems' Hawaii Laboratory, features an acrylic sphere; its name means "Eye in the Sea." The craft rests on a platform that can drop as far as 200 feet to launch and recover the vessel. Waves create less turbulence below the surface, permitting safer operation in choppy waters.*

NAVAL OCEAN SYSTEMS CENTER

DAN BUDNIK / WOODFIN CAMP AND ASSOC. (ABOVE); U. S. NAVY (BELOW)

GRUMMAN AEROSPACE

Ben Franklin, *named in honor of the statesman-scientist who first had the Gulf Stream mapped, looms above her designer, Jacques Piccard. Twenty feet high and nearly 50 feet long, the craft can remain 2,000 feet down for a month or more. After a 1969 test dive near Palm Beach, Florida, Dr. Piccard, wearing a red life jacket (below, far left), helps transfer a bag of equipment to a raft. On July 14, 1969, a six-man crew of scientists and engineers submerged (below) to begin an undersea study of the Gulf Stream from Florida almost to Nova Scotia. For 30 days they drifted with the current, studying marine life and recording such characteristics as water temperature and salinity. They also investigated the effects of isolation on a group working together—a study sponsored by the National Aeronautics and Space Administration. Upon completion of the 1,500-mile voyage, a reporter asked Dr. Piccard to describe what he had seen. "The most spectacular thing we saw," the scientist replied, "was probably the sunlight when we opened the hatch."*

N.G.S. PHOTOGRAPHER BATES LITTLEHALES (BELOW); RAGNAR M. PETERSEN / LOCKHEED

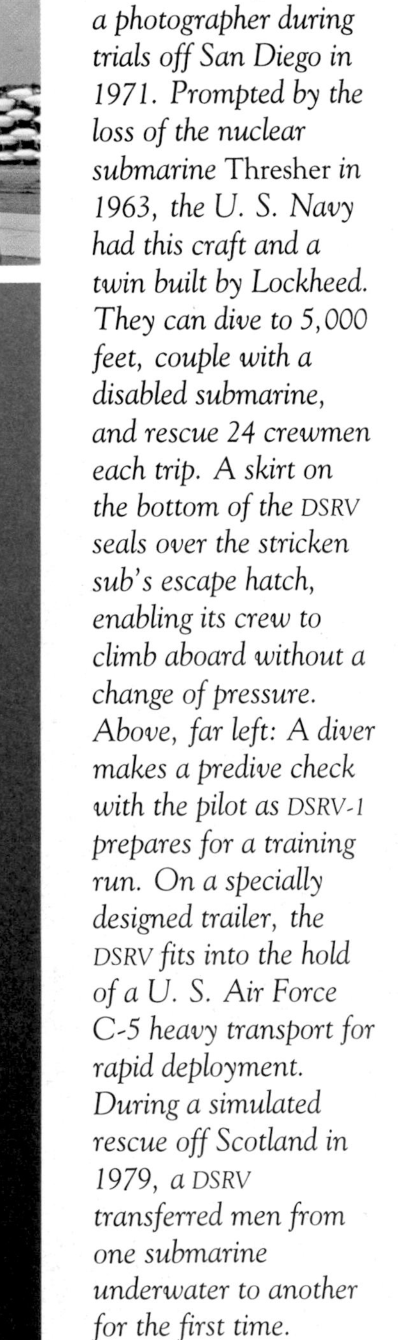

Rescue submarine DSRV-1 *glides past a photographer during trials off San Diego in 1971. Prompted by the loss of the nuclear submarine* Thresher *in 1963, the U. S. Navy had this craft and a twin built by Lockheed. They can dive to 5,000 feet, couple with a disabled submarine, and rescue 24 crewmen each trip. A skirt on the bottom of the* DSRV *seals over the stricken sub's escape hatch, enabling its crew to climb aboard without a change of pressure. Above, far left: A diver makes a predive check with the pilot as* DSRV-1 *prepares for a training run. On a specially designed trailer, the* DSRV *fits into the hold of a U. S. Air Force C-5 heavy transport for rapid deployment. During a simulated rescue off Scotland in 1979, a* DSRV *transferred men from one submarine underwater to another for the first time.*

N.G.S. PHOTOGRAPHER BATES LITTLEHALES (BELOW AND ABOVE, RIGHT); NOAA

Versatile and maneuverable, submersibles perform a variety of tasks considered impossible or impractical as recently as the 1950s. At left, divers make a last-minute inspection of Star III *as she prepares for a 1967 mission off Key West. Two 35-mm cameras mounted under her bow will take stereophotographs of the ocean floor, enabling cartographers to map it in accurate detail.* Nekton Beta *(above, left) rests gently near a diver collecting samples of sea life.* NOAA *has used the craft in studies of reef ecology and marine fisheries; in the 1970s it enabled scientists to explore East Coast fish and shellfish grounds and observe biological effects of ocean dumping. Above,* Asherah *searches for ancient shipwrecks off Turkey. Named for the Phoenician sea goddess and built for the University of Pennsylvania with grants from the National Science Foundation and the National Geographic Society,* Asherah *can map an undersea archaeological site in minutes—a task that would take divers weeks to complete.*

N.G.S. PHOTOGRAPHER EMORY KRISTOF (ABOVE AND BELOW, LEFT AND RIGHT)

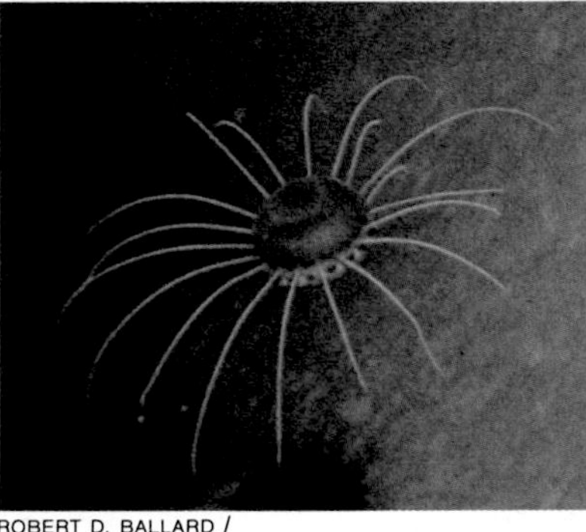

ROBERT D. BALLARD /
WOODS HOLE OCEANOGRAPHIC INSTITUTION

Cloud of silt settles about Alvin *as she carries scientists to the sloping walls of the Cayman Trough early in 1976. During the expedition, geologists for the first time could see and collect rock samples that revealed the composition of the deep-ocean crust. At left, a diver inspects* Alvin *before descent. Opposite, center, another diver checks the angle of a camera system on a sled before it plunges to pinpoint the best locations for* Alvin's *dives. Towed by a surface ship and operated by remote control, the sled holds battery-powered strobe lights and two cameras that take color pictures every 10 seconds. Seen through* Alvin's *view port, a jellyfish moves through the Cayman Trough (opposite, lower) by expanding and contracting its body.*

After a daylong dive into the Cayman Trough, Alvin *churns slowly toward* Lulu, *a catamaran tender. The sub's pilot stands in the sail; a diver hitches a ride. Crewmen waiting on the support ship fling lines to* Alvin, *then guide her into position between* Lulu's *pontoons. An elevator there lifts* Alvin *out of the water to the main deck for supplies and maintenance. Right, ready to tie down the sub's basket of instruments, a diver leaps toward* Alvin *as she surfaces. Another swimmer heads toward the vessel, carrying a telephone to talk with the men inside.*

NATIONAL GEOGRAPHIC PHOTOGRAPHER EMORY KRISTOF

NATIONAL GEOGRAPHIC PHOTOGRAPHER EMORY KRISTOF

Bright red worms encased in white tubes as long as eight feet cluster on the seafloor near a warm-water vent in the Galapagos Rift. The unexpected 1977 discovery of these oases of life prompted further investigation in 1979. Above, a diver makes a final inspection before Alvin *submerges. The sub carries a miniature color TV camera developed by* RCA *that provides close-ups of unusual clarity. Below, clockwise from upper right:* Alvin's *claw picks up a foot-long clam; brachyuran crabs crowd into a fish-baited trap. Far to the north on the East Pacific Rise, a vent spews mineral-rich water at more than 662°F (350°C).*

AL GIDDINGS / SEA FILMS, INC. (OPPOSITE)

JOHN EDMOND, M.I.T. AND NATIONAL SCIENCE FOUNDATION

ROBERT D. BALLARD / WOODS HOLE OCEANOGRAPHIC INSTITUTION

WOODS HOLE OCEANOGRAPHIC INSTITUTION

Amid Alvin's *array of switches, dials, and indicator lights, marine biologist Dr. J. Frederick Grassle settles back for the hour-and-a-half descent to the Galapagos Rift. "It was fantastic," he recalls of his first visit to the unique ecosystem. "It was like another world, a lost valley." At right, excited scientists gather around the first of the tube worms brought to the surface. Microbiologist Dr. Holger W. Jannasch examines a sample collector. He and others concluded that unlike most communities of life, which depend ultimately on the sun for existence, these deep-sea creatures depend on bacteria that derive their sustenance from minerals dissolved in the water, not from organic matter. Opposite, zoologist Dr. Carl J. Berg carefully dissects a mussel. Unusual coloring of a giant clam comes from hemoglobin, the iron-rich substance in human blood that carries oxygen to tissues. Such discoveries might never have taken place without the use of submersibles. Expensive to build and costly to operate, these vessels have just begun the exploration of the ocean floor.*

AL GIDDINGS / SEA FILMS, INC. (BELOW AND BOTTOM, RIGHT); N.G.S. PHOTOGRAPHER EMORY KRISTOF

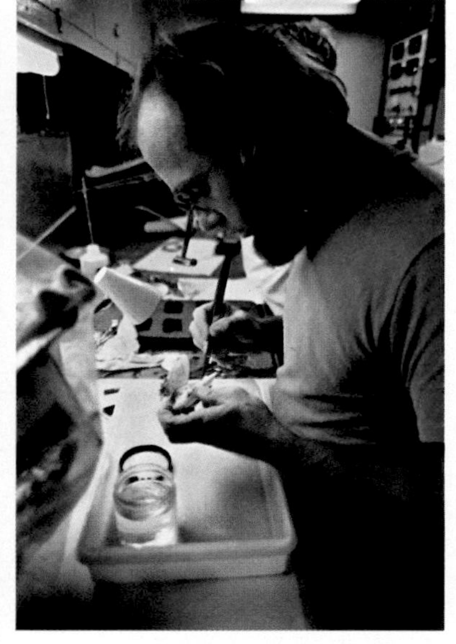

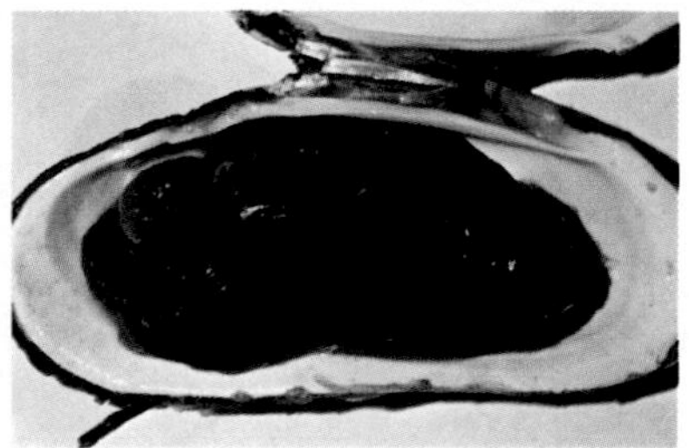

The Ocean's Wealth and Health

"WHAT IS THE USE of knowing how deep [the sea] is unless we know what is at the bottom of it?" questioned Matthew Fontaine Maury when charged with surveying the deep sea as superintendent of the U. S. Navy's Depot of Charts and Instruments in 1842. "Where was the mechanical skill," he continued, "that would contrive for us the means of bringing up from miles below . . . the feathers from old ocean's bed, be it ooze, or mud, or rock, or sand. . . ?"

Not until 1872 was a concerted effort made to explore worldwide oceans scientifically. Then H.M.S. *Challenger* set out from Portsmouth, England, to seek answers to fundamental questions such as: "How deep is the sea?" "What is it like on the ocean floor?" "Is there life in the great depths?"

Government instructions to *Challenger*'s captain read, ". . . you have been abundantly supplied with all the instruments and apparatus which modern science and practical experience have been able to suggest and devise . . . you have a wide field and virgin ground before you." There were no masks or flippers aboard *Challenger*, and the first underwater photograph would not be taken for two decades. But dredges, nets, trawls, and special bottom-sampling equipment brought ooze, mud, and rock to the surface and yielded thousands of plant and animal species new to science. The scientists aboard during the 3 1/2-year cruise were not aware that the strange mineral nodules they found—curiosities to them—would be coveted in future times, nor could they guess that a later generation would prize another resource from the ocean floor.

A hundred years later that valuable resource—oil—supported a thriving industry about 500 miles from *Challenger*'s home port. In 1979 I listened to Dick Clarke, saturation-diving supervisor for Oceaneering International's North Sea oil contracts, as he relived weeks spent in the heart of the Ekofisk fields in Norwegian waters: "The wind was howling at 40 knots. Waves 26 feet high smashed into the platform, drenching me with spray and spume. Just within our working limits, we lowered a diving bell to work 250 feet below." Dick and I had collaborated many times on Hydro-Lab projects near Freeport, Grand Bahama Island. I knew this man to be addicted to warm, clear, tropical seas, and the vision of him now spending much of his time on offshore North Sea oil rigs stretched my imagination. But Dick finds challenge at the cutting edge of diving action.

"Ninety percent of full-time commercial divers work with offshore oil," he told me. "Exploration, construction, and maintenance of offshore rigs have been powerful incentives in developing new diving techniques. What we're doing today compared with what was possible only 10 years ago is unbelievable." Though it seemed bizarre

"Nothing is great but the inexhaustible wealth of nature. She shows us only surfaces, but she is a million fathoms deep."
Ralph Waldo Emerson, "Resources"

Billows of rock cuttings and mud—drilling debris—envelop a U. S. Geological Survey diver near the outflow pipe of an oil-drilling platform off the coast of Texas. Water samples collected at various distances from the pipe help assess the effects of such pollution on undersea life. This project, one of many current marine studies, reflects increasing concern about the ocean.

STEPHEN C. EARLEY

during Tektite II in 1970 to live 50 feet underwater for 14 days, the innovative methods Dick now described take divers to work at 1,000 feet and keep them at that pressure for weeks at a time.

"Our contract calls for 'year-round diving capability,' and that means dealing with the worst elements of the North Sea, above water

VOYAGE OF H.M.S. CHALLENGER, SMITHSONIAN INSTITUTION

Pioneer in ocean research, H.M.S. Challenger *probes mid-Atlantic shallows off St. Paul's Rocks (right) during history's first global oceanographic expedition. Setting out in December 1872 with instructions to explore "all aspects of the deep-sea," the ship traveled 68,890 nautical miles through every ocean but the Arctic during her 3½-year voyage. On board: hundreds of sample bottles, miles of sounding line, and a laboratory containing the latest scientific instruments. Steam engines reeled in* Challenger's *deep-sea dredges and trawls with the aid of blocks and cylindrical spring devices mounted in the rigging (left). Though crude and cumbersome by today's standards, such equipment enabled* Challenger *to chart 140 million square miles of ocean floor; to gather information on currents, sediments, and temperatures; and to collect 4,417 new species of marine plants and animals.*

and below. We now launch and recover a diving bell with men inside in weather that once was considered impossible," continued Dick.

On a saturation dive the bell descends rapidly from the surface—120 feet a minute—down a line held taut by four tons of ballast on the seafloor. Once it is in place, the men inside open the hatch and swim out.

As Dick spoke, vivid images emerged of those who use diving not as an end in itself, but as a means of getting to where they have to go. Wearing a "hot water" suit and close-fitting helmet, a man leaves the warm, dry atmosphere of the bell and glides into bone-chilling water. His light slices through the murky sea, guiding him to an intricate steel maze at the base of the platform. Following like a slender serpent is a lifeline conveying four essentials: hot water that circulates through his suit, giving blessed warmth; a rigidly controlled breathing mix of oxygen, nitrogen, and helium; the reassuring voice of Dick Clarke—or another diving supervisor—who maintains continuous two-way communication from the surface control room; and a line to measure depth. The trailing tether also serves as a vital "come-home" line amid the web of look-alike steel branches.

The diver carries tools for the tasks he has come to do, tasks usually similar to those done on terrestrial drilling rigs, such as inspection,

welding, replacing worn parts, repairing broken sections, and laying new cable or pipe. A veteran diver advises commercial diving students to become expert at these jobs, as well as at special underwater operations. "A diver is not paid because he can go underwater," he notes. "A diver is paid for the work he performs. . . ." The pay is excellent, but the work is hard and the risk high. From 1971 through 1979, 45 divers died in the North Sea in oil-related accidents.

Divers especially have to learn to combat panic. Even a momentary distraction, trivial on land, can have disastrous consequences 500 feet underwater. But with caution and adherence to safety guidelines, work on oil rigs can be performed safely. Dick Clarke told me that in 1978—a year in which divers from the Ekofisk complex performed 2,952 hours of work on the bottom—the most serious health problems and accidents were injured fingers and ear or skin infections, including five cases of athlete's foot acquired in the platform decompression chamber!

In the North Sea four-hour dives in water ranging from 39° to 50°F (4° to 10°C) outside a bell are routine. After one diver returns, a second usually goes out for another four hours of work before the bell is raised to the surface and mated with a large deck decompression chamber. There the divers, although still experiencing the same pressure as

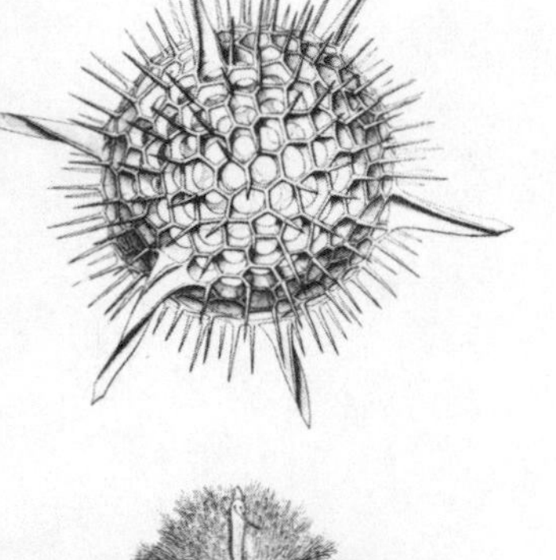

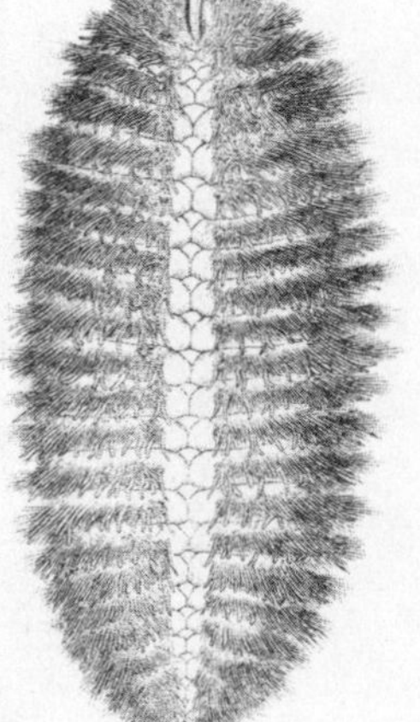

at the underwater work site, have more space in which to move around and have access to a hot shower, comfortable bunks, dry clothes, and—on most oil rigs—excellent cuisine. After some rest the cycle is repeated for three or four weeks until the entire saturation complex is decompressed or the divers are replaced.

The deepest working saturation dive ever made took place in October 1977, when divers from the French company COMEX completed a pipeline hookup at 1,510 feet. COMEX also carried out two simulated dry-chamber dives at 2,001 feet in 1972 and 1974. Generally, however, work on offshore oil rigs takes place at 300 feet or less, and much of it is done by divers who descend from the surface singly or in teams.

Diverse menagerie of sea life parades across the pages of Challenger's *journals. The creatures shown here represent a tiny fraction of thousands collected, examined, and classified by the ship's four naturalists. They include (clockwise, from above): a cigar-shaped segmented marine worm; a spherical protozoan known as* Hexastylus;

TODAY the entire continental shelf area of the world—the region that extends from shore to about 600 feet down—is accessible to divers. We are close to realizing what Edwin A. Link hoped to accomplish when he began his Man-in-Sea Project and described the exploration of the shelf in 1963: "If man could find a way to work there in safety and relative comfort," Link wrote, "he would at once possess the key to more than 10,000,000 square miles of sea bed. He could tap the scientific secrets and mineral, animal, and vegetable wealth of these immense submerged plains. . . ."

But in fact, although the key is now in hand, diving technology is not widely applied. Instead, because of economic pressures, the most up-to-date equipment is used in financially rewarding situations—particularly in the oil industry. In 1978 about 20 percent of the world's oil supply came from the sea; the percentage will undoubtedly increase.

According to geologist Dr. Kenneth O. Emery of Woods Hole Oceanographic Institution, the amount of oil underwater far exceeds that remaining on land. Even the oil found in deserts had its origin in ancient seas, where it derived from the remains of marine plants and animals. Offshore, along the Atlantic coast of the United States, there may be as much as four billion barrels of oil and 17.5 trillion cubic feet of natural gas. The demand for energy in the United States alone—78.4 quadrillion BTU's in 1978, of which oil contributed near-

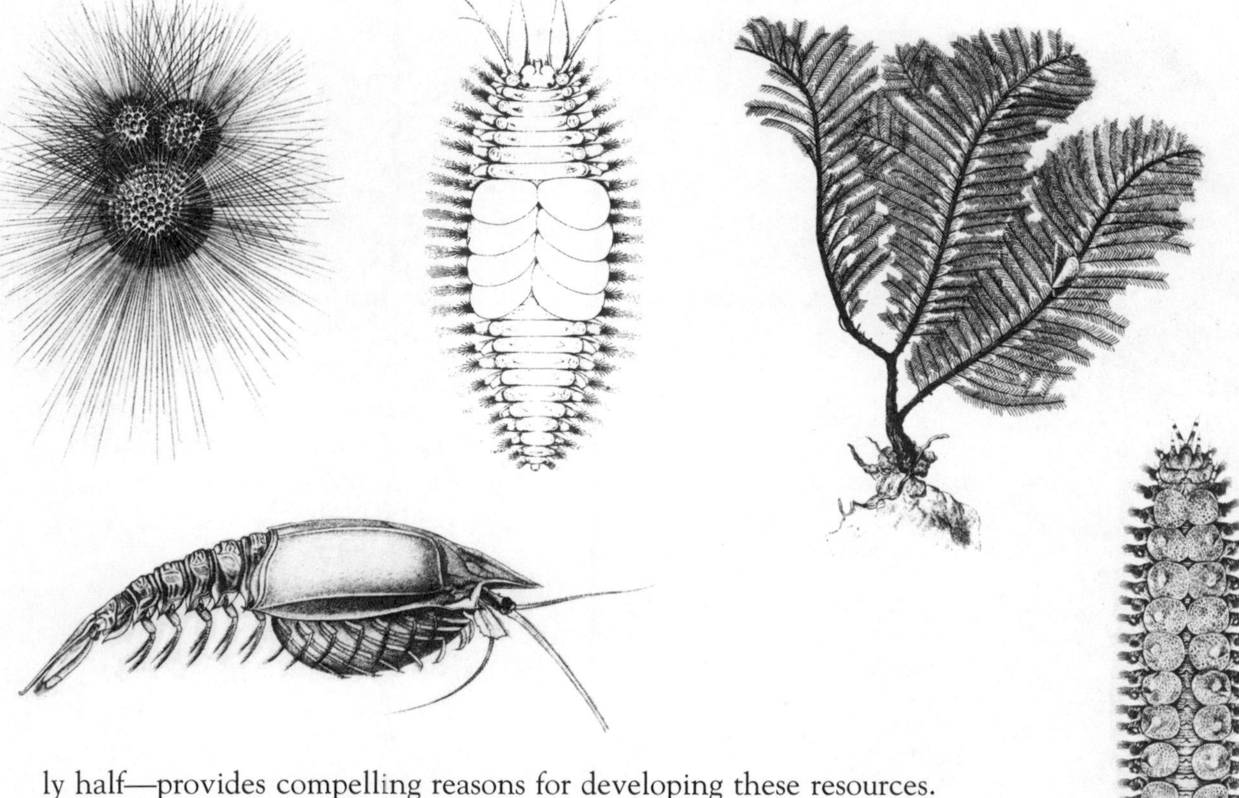

ly half—provides compelling reasons for developing these resources.

Economically, offshore oil and gas are of such importance that they overshadow all other sea resources. According to Dr. Emery, the total value of minerals exploited on the continental shelves worldwide in 1977—mostly gas and oil—was 70 billion dollars. That was more than three times the dollar value of animal resources taken.

But there are trade-offs for the drilling of oil. The world's largest recorded oil spill occurred in the summer of 1979 when drilling in the Gulf of Mexico resulted in a massive blowout. According to the Center for Short-Lived Phenomena, more than 450,000 tons of oil gushed into the sea. The oil drifted northward, some of it washing up on Texas beaches. By early 1980 various efforts to combat the blowout had reduced the flow to about 280 tons a day.

Wells are not the only source of trouble. Even transporting the oil can be dangerous. In 1967 the tanker *Torrey Canyon* broke up off Cornwall, England, spilling 119,000 tons of crude oil onto beaches and turning the sea, for a while, into a thick, dark pudding. Studies continue to assess the impact and provide guidelines for avoiding such disasters in the future. But similar episodes have followed.

The *Argo Merchant* ran aground off New England in 1976, pouring 25,000 tons of oil into the sea; and in 1978, in one of the most dramatic events in supertanker history, the *Amoco Cadiz* lost its steering and struck rocks near Portsall, France, dumping 223,000 tons of Arabian oil over some 2,000 acres of oyster beds and throughout Brittany's rich fishing grounds. More than 3,000 dead birds littered the beaches. Local residents who traditionally lived by fishing now faced an uncertain future. In protest they posted hand-printed signs that read, "*La mer est morte*"—the sea is dead.

Such catastrophes are awesome, but the accumulation of small

side and head views of a dragonfish found in Antarctic waters; three round, spined planktonic animals called Globigerina; *another sea worm; feathery fronds of a Pacific coral; yet another marine worm; a deep-sea crustacean related to the shrimp; and a toothy bottom-dwelling fish from the Mediterranean Sea.*

VOYAGE OF H.M.S. CHALLENGER, SMITHSONIAN INSTITUTION

spills also can be dangerous. Because of legislation that requires the pumping of oily water into containers for safe disposal in port, dumping of such waste by ships at sea has been greatly reduced during the 1970s. But tank washing still accounts for at least one-sixth of the six million tons of oil that enter the ocean through human activity each year. Some ships continue to spill patches of brown, tarry material. And if you are only 18 inches high, like the black-and-white jackass penguins that inhabit the waters along the coast of South Africa, a yard-wide puddle of oily water can be as deadly to you as a million gallons of crude oil floating on the sea.

SYLVIA A. EARLE

With Jacques Verster, Director of the Southern Africa Nature Foundation, and biologists Rod and Bridget Randall, I visited the newly established island sanctuary at St. Croix in July 1979. The Randalls have lived at this South African rookery periodically for three years, studying its raucous penguins and other birds.

"This one looks healthy," Rod indicated, pointing to an adult female streaked with splotches of brown oil. She hovered over two small gray puffs, young penguins.

"In fact, she is doomed, as are the chicks. Her feathers are too stuck together to provide adequate insulation when she goes to sea for food. Without her protection, the chicks will die. It has already happened to hundreds of others."

But oil-drenched penguins are more fortunate than flying seabirds; some of the penguins can be rehabilitated. Stricken birds are flown to Cape Town, where the South African Foundation for the Conservation of Coastal Birds cleans, feeds, and releases them. In July 1979, 150 penguins from St. Croix were treated in Cape Town, approximately 550 miles away by sea. Says Bridget, "Our recovered penguins are returning. We now have 41, and their shortest time finding their way back to St. Croix is an amazing 13 days."

In recent years research institutions, government, and industry have begun extensive programs to try to establish the effects of oil spills on marine life and to explore ways of restoring damaged areas. The concern about the lack of adequate information prompted the U. S. Bureau of Land Management in 1974 to gather appropriate data about the nation's continental shelves at key locations. Although some people hope that within 20 years alternate energy sources will sharply reduce the demand for fossil fuels, for the time being, at least, the pace of finding and taking oil from the sea seems unlikely to slacken. And if there are problems in the wake of continued offshore drilling, at least there will be documentation of where things stood in the mid-1970s.

Meanwhile, several nations are planning to mine other potential wealth from the sea—in the form of dark, potato-shaped mineral nodules that could be worth trillions of dollars. An estimated 90 billion to 1.7 trillion tons of these nodules, first discovered during the historic H.M.S. *Challenger* expedition in the 1870s, are believed to lie on ocean floors. They are concentrated in the deeper parts of the Pacific basin on sediments that are accumulating very slowly—only a few millimeters every thousand years. The nodules could contain 10 times the land reserves of manganese, 13 times those of nickel, 1.3 times those of copper, as well as large amounts of cobalt.

The *Deepsea Miner II* is a ship capable of reaching through 15,000 feet of water to retrieve the nodules with an airlift system. The *Glomar Explorer* can retrieve nodules too, but the Deep Sea Drilling Project, which has supervised the operation of the *Glomar Challenger* for 12 years, hopes to convert the *Explorer* for research drilling in the 1980s. Equipped with pinpoint navigation and machinery that enables it to remain stable in seas that average 20 feet, *Explorer* can manipulate more than 30,000 feet of drill pipe.

I stood on the edge of *Explorer*'s 200-foot-long rectangular well, through which machinery and a pipe string are manipulated, and pondered our access to the ocean. "It is difficult for most divers to go as deep as the opening is long," I thought. "Yet that opening can handle more than five miles of pipe to probe far below the seafloor—and bring sediment to the surface."

The man who showed me around *Explorer* handed me a walnut-size nodule that encased a shark's tooth. "That piece is about 10 million years old," he said. "Minerals are added in thin layers, like those of an onion."

The thought startled me. Manganese nodules were first discovered only about a hundred years ago. To most of us a century is a long time, but geologically speaking it is no time at all. Most recent estimates suggest that a Pacific deep-sea nodule grows at the rate of about three to six millimeters each million years.

Nodules are formed around bits of rock, sharks' teeth, and sometimes even the beak of a squid. Manganese and iron are thought to precipitate from seawater just above the sediments. The most valuable nodules are those especially rich in nickel and copper, and these seem to form in areas where a heavy rain of dead animals and plants brings high concentrations of these metals to the bottom. Living organisms such as bacteria also may influence the development and composition of these deep-sea nodules.

The ocean floor is regarded by some experts as a relatively lifeless place, but recent evidence indicates that nodules found here stay free of sediment and do not become buried readily, possibly because of the activity of associated organisms. Other deep-sea areas, places once thought to be undersea "deserts," are now known to be as rich in marine life, inch for inch, as tropical shallow-water systems.

Some influential people are trying to keep these discoveries in perspective as plans to mine the seafloor accelerate. Early in 1979 Robert Wicklund, a member of Senator Lowell Weicker's staff, helped to frame Senate Bill 493, "to promote the orderly development of hard mineral resources of the deep sea bed. . . ." Wicklund, for five years the manager of the Hydro-Lab project in the Bahamas, has recently translated his working knowledge of the ocean into efforts to develop responsible ocean policies.

"I miss being part of the sea," Bob told me, "and someday I'll get back. But for now, it seems more important to use my experience to influence responsible oceans legislation."

If passed, the Deep Seabed Mineral Resources Act would be a legal model. The bill sets guidelines for assessing the impact of mining on the ocean floor and sets up protection for the investments of firms already engaged in mining. Even before the bill was introduced, one U. S. company had laid claim to an area of the eastern Pacific and constructed a vessel for test mining on a huge scale. Because of the tremendous costs involved, as much as one million tons of nodules a year would have to be processed to make the undertaking profitable.

The Senate bill, while applicable only to U. S. citizens, may set significant precedents for regulating the removal of deep-sea resources. Aside from providing for obvious benefits—recovery of the

Captured on paper by Challenger *scientists in the 1870s, jackass penguins now live a precarious existence on island rookeries off Africa's southern tip. Wrote one scientist on the historic oceanographic voyage: "All the birds fought [us] furiously. . . . They make noise very like the braying of donkeys, hence their name; they do not hop, but run or waddle." Man has not shown kindness toward these noisy birds, which numbered 1.5 million in the 1920s but may number about 50,000 today. Once heavily exploited for their eggs, they now must compete for food with South Africa's growing inshore fisheries. Tankers pose yet another threat: Bilge-oil pollution mats the birds' feathers (opposite), robbing them of their insulation. Jacques Verster, Director of the Southern Africa Nature Foundation, seeks out begrimed victims for rescue and cleaning.*

VOYAGE OF H.M.S. CHALLENGER, SMITHSONIAN INSTITUTION

Undersea elevator lifts U. S. Navy divers toward the surface. Periodic stops allow them to decompress during tests of the innovative Mark XII diving system. Far lighter and more efficient than previous hard-hat gear, the Mark XII increases a diver's mobility and comfort and permits routine diving to 380 feet. A closed-circuit television camera mounted on the helmet of the diver at left allows technicians topside to monitor the equipment's performance.

BERNIE CAMPOLI / U. S. NAVY

minerals—the bill should spur debate on such unresolved questions as these: What effect will mining have on deep-sea life or on circulation of water? How would changes in the sea affect life on dry land?

While lawmakers debate, others grapple with the haunting unknowns that make rational policies concerning the law of the sea so difficult to arrive at. Among the seekers are scientists working from a 20th-century H.M.S. *Challenger,* the ocean-drilling research vessel *Glomar Challenger.* Since its launching in 1968, the ship has retrieved more than 208,900 feet of core samples from more than 500 sites around the world, providing scientists with the raw material for some intriguing findings. Basing their conclusions on the columns of hardened sediment and cooled lava raised by today's *Challenger,* scientists have amassed new information about the creation of the polar ice caps, worldwide volcanic activity, changes in the earth's magnetic field, the rising and falling of land masses, the evaporation of the Mediterranean Sea, the evolution and extinction of sea life, and the possible location of deep-sea oil fields and various mineral deposits.

Perhaps most important have been *Glomar Challenger*'s spectacular contributions to the theories of seafloor spreading and plate tectonics. Samuel W. Matthews, writing in a 1973 NATIONAL GEOGRAPHIC report about revolutionary changes in geological thinking, observed: ". . . the most convincing evidence of all came home in mud and rock drilled from the floors of the world's oceans by a gangly, improbable seagoing drill tower named *Glomar Challenger.* . . . designed . . . by Global Marine, Inc., to lower more than 20,000 feet of pipe in the open ocean, bore into the sea floor, and bring up bottom cores or samples. The technical feat has been likened to drilling a hole in a New York sidewalk with a strand of spaghetti dangled from the top of the Empire State Building."

The pattern of samples brought up by the ship indicates that the earth's crust consists of huge plates moving slowly upon the globe. Volcanic action pushes new core material up along midoceanic ridges; old crust disappears into the deep-ocean trenches where two plates meet. One result of this activity is that the Atlantic Ocean is widening by one to two inches a year.

TANGIBLE REWARDS come from today's sophisticated explorations, whether undersea or in outer space. The *Glomar Challenger* and the Apollo lunar module Challenger with which it shared a name retrieved core samples and moon rocks that will keep researchers busy for years. But the intangible rewards are greater: a perspective that helps explain the past and offers direction for the future.

Now that we have glimpsed the significance of the deep sea, our attitude toward that part of the ocean should nevermore be one of complacency. Though we burn fossil fuels and though we are about to process millions of years of geologic history in recovering valuable minerals from nodules, we must protect enough of the ocean to ensure the continued stability of the living systems there.

Professor Jacques Piccard suggested in 1978 that the challenges of the past had included attaining great heights and plunging to great depths, as well as exploiting the world's resources above water and below. "However, this is not the last challenge," Piccard observed; "a new one, much more exciting, is the one which concerns the preservation of the sea. Blindly exploiting the sea," he warned, "may result in the destruction of a great part of it and in the modification, maybe for thousands of generations, of its environment."

IRA BLOCK (TOP AND BOTTOM)

Submersible shuttles resemble hardware used in the U. S. space program. Marvin Poole (above) communicates with a surface control room from the service capsule he pilots between a support ship and seafloor oil wells in the Gulf of Mexico. Tended by a cable and equipped with an umbilical line that carries air, electricity, exhaust hoses, and communication signals, the capsule

IRA BLOCK / WOODFIN CAMP AND ASSOC.

(top, left) descends into the deep. There it will dock with a larger, air-filled "cellar" capping the actual wellhead. Operators aboard the capsule will enter the cellar (left) to repair or adjust fittings that direct the crude through undersea pipes to a production platform. The system—developed by Can-Ocean Resources, Ltd.—avoids the expenses and dangers of maintaining such giant offshore platforms as Phillips Petroleum's Ekofisk complex in the North Sea (PRECEDING PAGES). The huge steel structures rise 400 feet from the continental shelf, supporting machinery for pumping oil and gas, helipads, and living quarters for more than 200 people.

N.G.S. PHOTOGRAPHER EMORY KRISTOF (PRECEDING PAGES)

TED JAYNES / INTERSUB

Commuting to work aboard a special mother ship, one of a new breed of submersibles (above) proves its worth in speeding construction and simplifying maintenance of offshore facilities. Arriving at this station, the vessel cuts loose and dives as deep as 1,000 feet. Self-contained and battery-operated, it roves free of restrictive cables and umbilicals.

REMY DURAND / INTERSUB

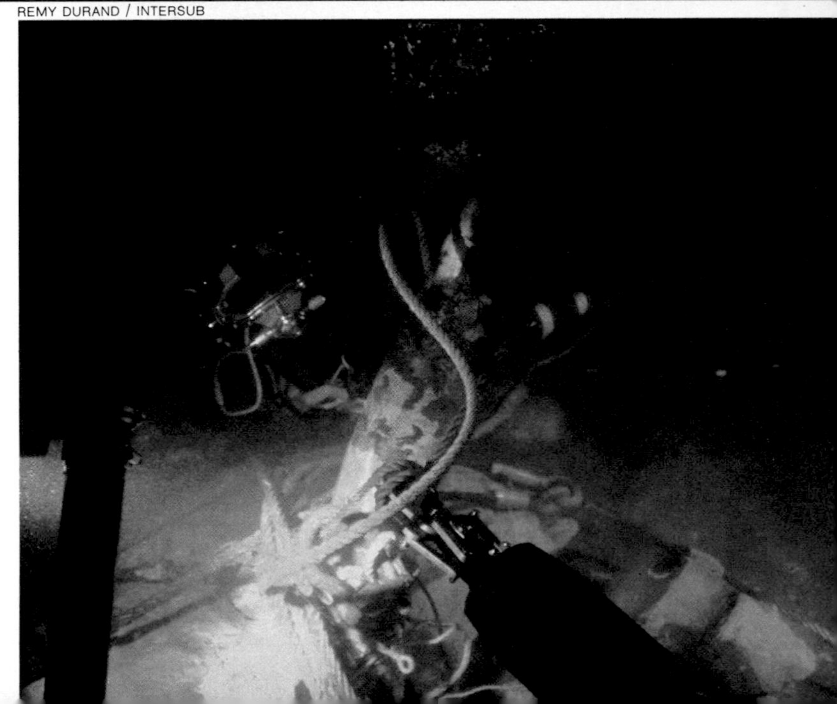

Large, bow-mounted observation port (below) gives the crew of a sub its main window on the sea; mechanical arms enable the workers inside to perform some underwater jobs in a shirt-sleeve environment. Certain models contain diver lock-out ports through which divers can leave and reenter the vessel. Deep in the Strait of Messina (bottom), a sub crew oversees placement of a natural gas pipeline, propping up stretches of rigid pipe that lie over hollows in the uneven seafloor. The mechanical arm of another sub (opposite, bottom) helps a diver free a cable snarled on a North Sea oil rig.

INTERSUB

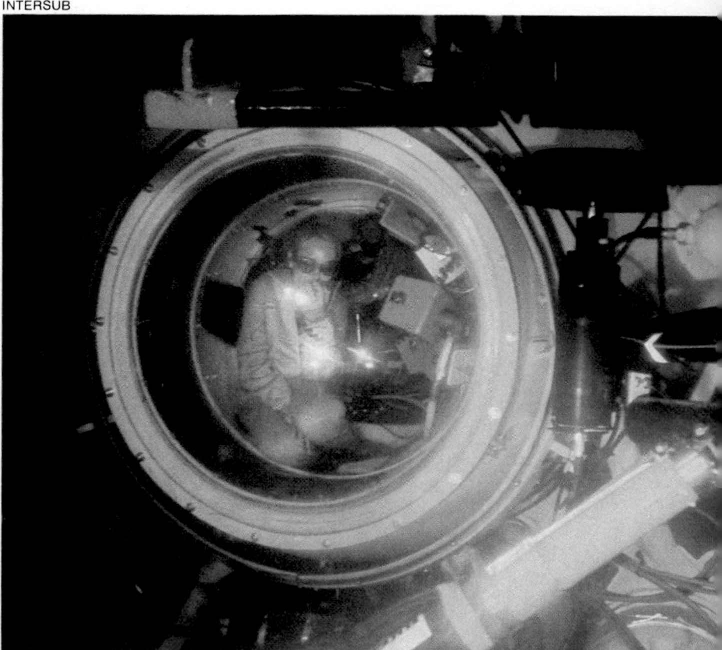

YVES OMER / INTERSUB

GLOMAR CHALLENGER
26
25
24
23
22

ALBERT MOLDVAY

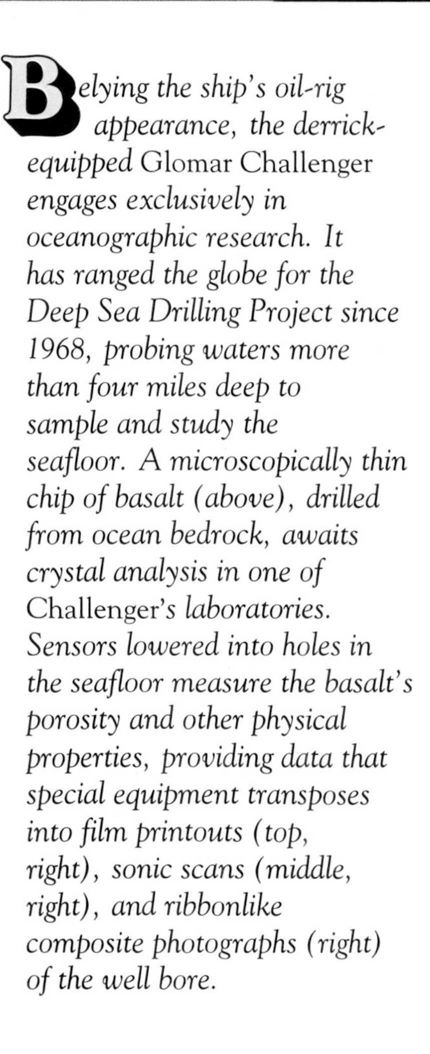

Belying the ship's oil-rig appearance, the derrick-equipped Glomar Challenger *engages exclusively in oceanographic research. It has ranged the globe for the Deep Sea Drilling Project since 1968, probing waters more than four miles deep to sample and study the seafloor. A microscopically thin chip of basalt (above), drilled from ocean bedrock, awaits crystal analysis in one of* Challenger's *laboratories. Sensors lowered into holes in the seafloor measure the basalt's porosity and other physical properties, providing data that special equipment transposes into film printouts (top, right), sonic scans (middle, right), and ribbonlike composite photographs (right) of the well bore.*

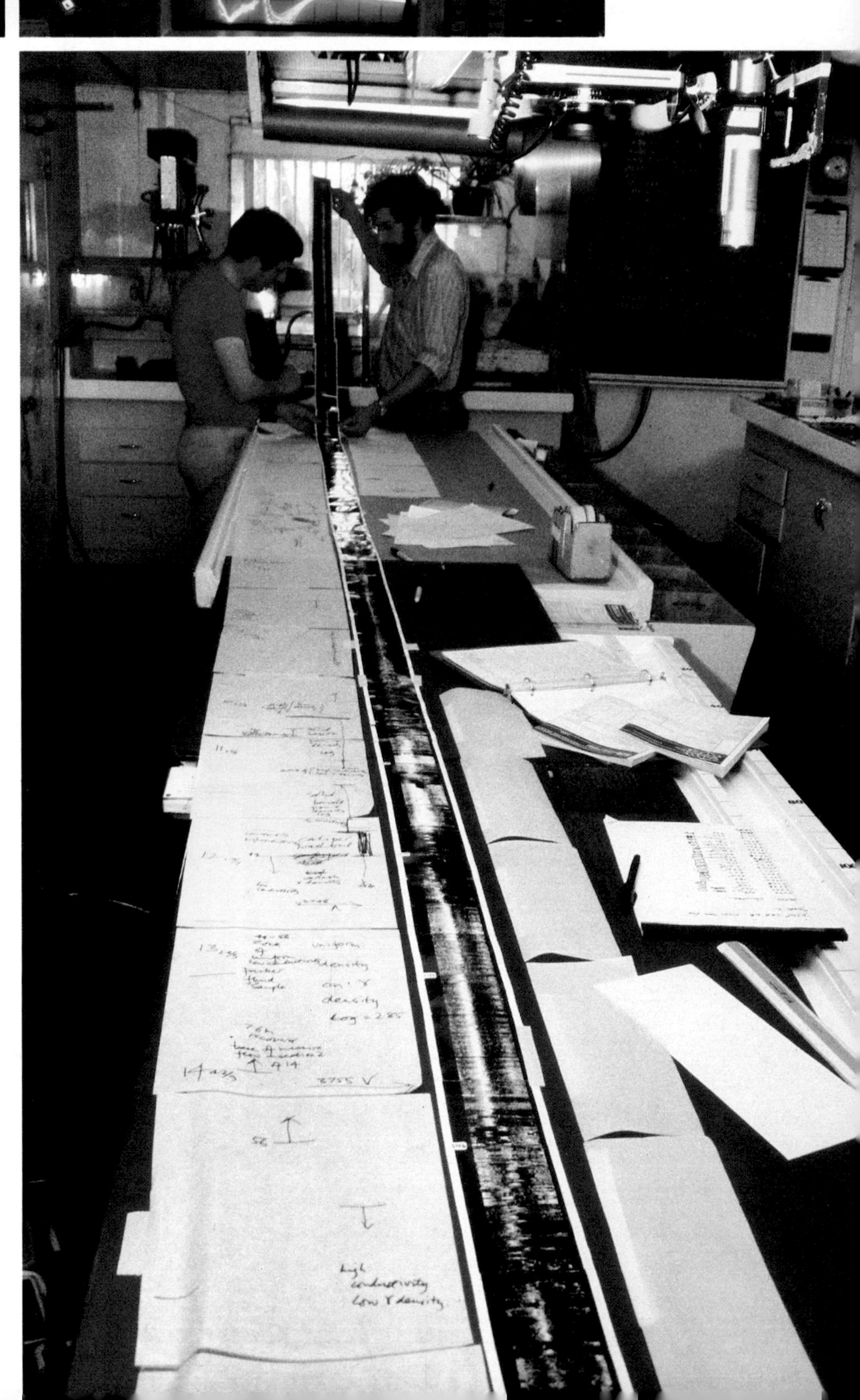

Aloft in Glomar Challenger's *steel rigging, derrickman Bill Needham (right) examines a connection on the flexible drill string—a hollow shaft that runs from ship to seafloor and turns the drill bit. As the drill bites deeper, heavy machinery hoists 90-foot lengths of pipe from a deck rack to the derrick and adds them to the string. The string then passes through an opening 20 feet in diameter in the center of the ship's hull (bottom). Computer-controlled thrusters allow the ship to maintain its position; sonic beacons and a huge steel funnel enable the drill string to reenter the hole after leaving it, in a rough equivalent of threading a needle's eye from 10,000 feet away—without seeing the needle.*

ALBERT MOLDVAY

Following a successful drilling, roughnecks (above) wrestle with pipe used to retrieve rock cores bored from the ocean bottom.

Breaking new ground in what may one day become a major industry, Deepsea Miner II *(below) tests ways to take manganese nodules from the ocean bottom. A gimbaled derrick rides amidships, capped by a weathertight geodesic housing. In addition to a pipe string, the derrick supports a dredge head (shown opposite during launching). Towed across the seafloor, it separates nodules from the sediments; compressed air then forces seawater and nodules up through the pipe string. During the operation an automatic plotter serves as the crew's eyes, continuously monitoring the positions of ship and dredge (bottom, left). Containing not only manganese, but also copper, cobalt, nickel, and other metals, the nodules litter certain areas of the ocean floor like pebbles on a beach (bottom, right). They form over millions of years, precipitating out of seawater like rock candy crystallizing from a sugar solution.*

E. DON BLANKENSHIP / DEEPSEA VENTURES, INC.

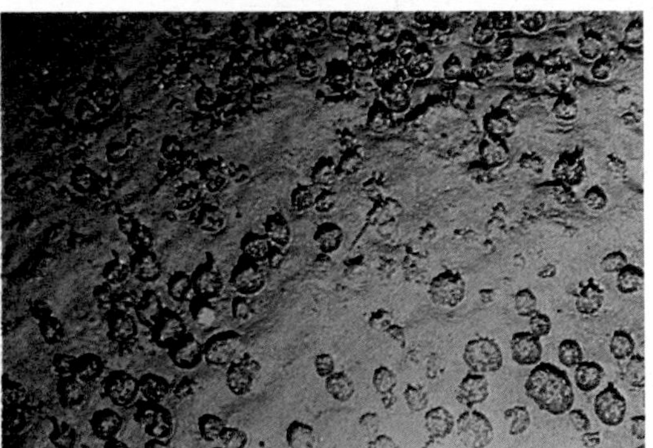

B. J. NIXON / DEEPSEA VENTURES, INC. (ABOVE)

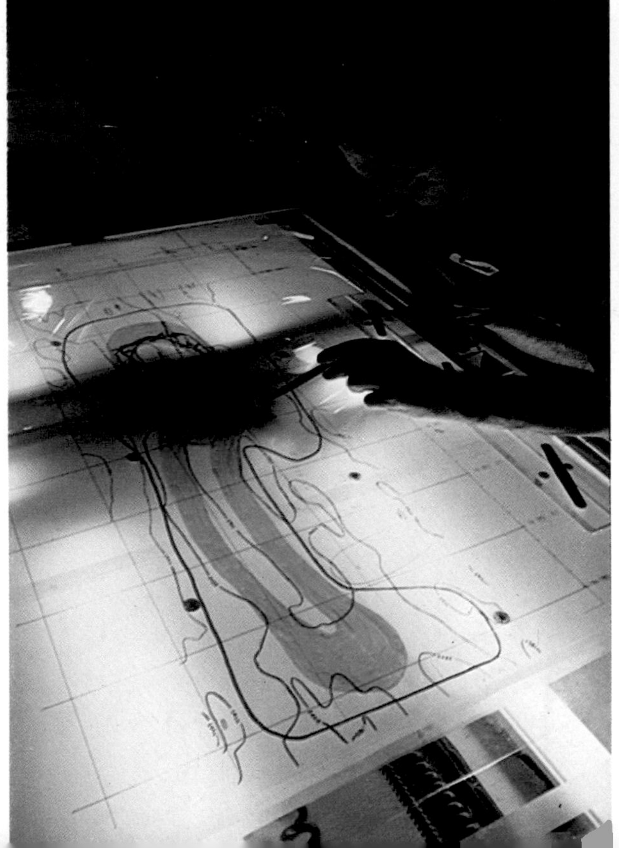

Too bulky for conventional ports, the 483,404-ton Globtik Tokyo*—one of the world's largest oil tankers—offloads its crude at a deepwater facility near Kiire, Japan. The ship boasts sophisticated navigational equipment, an inert-gas system to prevent explosions, and a*

tank-washing system that does not discharge oil into the sea. A new Globtik proposal: Construct a fleet of smaller but sturdier tankers that could smash through the ice-crusted Arctic Ocean and carry Alaskan crude from Prudhoe Bay to America's oil-hungry eastern coast.

MARTIN ROGERS

SIPA PRESS / BLACK STAR

MARTIN ROGERS (ABOVE AND BELOW)

Devastating oil tanker disaster hits Brittany in 1978 as the Amoco Cadiz *runs aground off the French town of Portsall and belches forth some 223,000 tons of Arabian crude. Winds and high spring tides drove the oil landward, drowning nearly 200 miles of shoreline in a chocolaty muck lethal to limpets (left) and other marine life. Thousands of seabirds died; oyster, lobster, and kelp industries foundered, as did the area's other mainstay—tourism. A 110-million-dollar mopping-up operation cleaned beaches (right), but long-term effects—especially on groundwater and the seafloor—will take years to assess.*

GUILLERMO ALDANA E. (ABOVE AND BELOW)

*Blowout! The sea itself seems to burn as natural gas from a shattered offshore wellhead boils up through oil-stained waters to feed surface flames on Campeche Bay in the Gulf of Mexico. Source of the world's biggest oil spill, this Mexican well—known as Ixtoc I—gushed from June 3, 1979, into 1980, at first pouring an estimated 4,000 tons or more of crude into the Gulf each day. Repeated attempts to cap the flow slowed but could not stop it; Ixtoc's growing slick drifted hundreds of miles north to Texas beaches, threatening the region's environment. Charges of mismanagement flew as oilmen, environmentalists, and national leaders debated what to do about the spill, fueling political and legal brushfires almost as hot as Ixtoc's flames (left). Fortunately, most oil platforms in the Gulf remain more serene (*FOLLOWING PAGES*). A recent U. S. Geological Survey study shows that between 1971 and 1978 only one blowout occurred for every 250 wells drilled in U. S. waters.*

STEPHEN C. EARLEY (FOLLOWING PAGES)

Seeking guideposts in an ocean of unknowns, scientists probe American waters for evidence of man's damaging effects. On Tanner Bank, 115 miles off San Diego, a diver (right) photographs pink hydrocoral. Aquarium-dwelling corals near Key Largo (below) receive various doses of drilling "mud"—a mix of clays, starches, and chemical additives used by the oil industry—in a test of their ability to survive pollution caused by offshore wells. USGS scientists (opposite) bore test holes in a Florida reef in an attempt to determine its age and geological history. Such studies help people exploit the wealth of the ocean without further endangering its health.

FLIP NICKLIN

STEPHEN C. EARLEY (BELOW AND RIGHT)

I II III IV
4 3 2 1

Reaching for the Limits

"We shall not cease
from exploration
And the end of all
our exploring
Will be to arrive where
we started
And know the place
for the first time."

T. S. Eliot, *Four Quartets*

ABOVE, BELOW, ON ALL SIDES, I was immersed in a translucent blue world, indescribable in commonplace terms. Cerulean? Sapphire? All are inadequate when I try to tell someone what open-ocean, deepwater blue is like. It seems to glow with its own light.

The blueness became distant, a dome over my head that shaded to blue-gray, then blue-black. Shadowy patches took form, and I felt a soft thump as the research submersible *Star II* gently touched the seafloor more than a thousand feet from the surface, six miles offshore from Makapuu Point in the Hawaiian Islands.

"Hold on," said the familiar voice of Al Giddings. "Boh is going to move on to look for deeper water." Bohdan Bartko, the lean, craggy-featured veteran of more than 500 dives as pilot of *Star II,* eased the sub along as I looked intently through the acrylic port of Jim. I was inside Oceaneering International's one-atmosphere diving system, strapped to *Star II* by means of a wide belt around Jim and standing on a platform attached to the sub's bow.

An 18-inch-long shark, dark above, pale below, moved gracefully across our path, its luminous green eyes momentarily turning in our direction. We passed through a field of bizarre bamboo coral, unbranched spirals that sprouted like giant, curling bedsprings from the seafloor. When touched at their tips, the corals pulsed with eerie blue light moving downward in glowing rings, and moving upward if the coral was gently nudged near the base. Amazingly, the light show lasted several minutes, the rings crossing and recrossing. I would have been content to spend all of my dive getting acquainted with these creatures, but we first had to find an appropriate site in deeper water where I could step off the platform and onto the seafloor.

Jim, a modern descendant of an early armored suit, was pressure-rated to 2,000 feet but had not been used deeper than 1,200 feet, except during the recovery of a TV cable off the coast of Spain. There the maximum depth Jim reached was 1,440 feet. We had anticipated no problem going to 1,500 feet to demonstrate further Jim's capability and to show the ease with which a diver in a one-atmosphere system could go to great depths, do a job, and return without the decompression penalty necessary for those exposed to outside pressure.

One of the first armored suits was designed in the early 1700s by John Lethbridge of England, who developed a wooden cylinder with armholes and a window for viewing. It was used for half-hour dives to 60 feet. In the 18th and 19th centuries many plans for such armored suits followed, but none could operate successfully below a few hundred feet. Great strides were made in the early 20th century through the engineering genius of a British inventor, Joseph (Pop) Peress. His first suit, built in 1924 of stainless steel, nearly immobilized the diver

Countdown to explore the cosmos of the deep: Inside a space-age diving system called Jim, author Sylvia Earle prepares to descend 1,250 feet to the Pacific Ocean floor six miles off Oahu for a solo walk. At the control panel of the Launch Recovery Transport pilot Troy Goodman makes final adjustments before releasing Jim from a support rack. Once off the LRT *60 feet undersea, the submersible* Star II *carried Jim to the depths, where Dr. Earle pioneered the suit as a tool for open-sea scientific research.*

AL GIDDINGS / SEA FILMS, INC.

within it by its sheer weight. But Peress had found the secret of successful operation at great depth. He used oil in the universal joints, thus reducing the likelihood that the arm and leg joints would lock under increasing pressure. His second suit, the "Iron Man," was made mostly of magnesium alloy and weighed about 800 pounds; it was used successfully in locating the sunken *Lusitania* in 1935. At that point the suit was evaluated by the Royal Navy and found "to come up to the claims of the inventor but no use could be made of it as the Navy [had] no requirements for deep diving."

LIBRARY OF CONGRESS

ARMORED DIVING was quietly shelved for 30 years while other methods of probing the sea were developed. But by 1968 going deep and returning without decompression had finally become desirable, especially for oil exploration and recovery. Peress, then in his 70s, was encouraged to dust off the old Iron Man and become a consultant to the developers of the modern Jims, named for the original Iron Man diver. Jim-9, the suit I was wearing, was equipped with a self-contained supply of air that was chemically scrubbed of carbon dioxide and recycled.

For weeks during the summer of 1979, Al had discussed with Phil Nuytten ways Jim could be used for scientific exploration in the open sea. Phil is an executive with Oceaneering International, Inc., which owns the suit and leases it and its counterparts for undersea work. A creative engineer himself, Phil combines an impressive record of commercial diving experience and a disarming sense of humor with the skills of an artist and the talents of a professional magician. All of these attributes were tapped many times during our collaboration.

Oceaneering approved a proposal that Al and I submitted, and we gained further support from the National Geographic Society, the Ocean Trust Foundation, the University of Hawaii, Maui Divers, and the California Academy of Sciences. At first Al and Phil planned to operate Jim in the traditional way: lowering the suit—with me inside—by a tether from a surface ship. Anchoring problems stimulated them to devise another approach. A transport belonging to Maui Divers would descend to 60 feet bearing Jim and the University of Hawaii's submersible, *Star II.* Then suit and sub would descend to the seafloor. It would be the first dive in a Jim suit ever made without a line to the surface, and the deepest solo exploratory dive in the open sea.

Although Jim was designed for an average-size male, not for a diver with my 110-pound build, Phil was confident that I could manage, once weights were added to Jim and properly balanced. A week of training at the Commercial Diving Center in Wilmington, California, convinced Phil and my trainer, William Turner, that I could walk, turn, lie down, get up, and even manage a modest, slow-motion cha-cha. I could manipulate Jim's two claws or leave my hands free for note-taking. It was no more difficult, mechanically, than learning how to row a boat or ride a bicycle, and no more taxing physically.

In predive exercises Phil and Al and I had gone through endless "what-if's": "What if *Star II* becomes tangled on the seafloor and can't go to the surface with you attached?" Al asked.

"I'll cut loose, drop weights," I replied, "and go up on my own."

"And what if *you* get tangled in something, and the sub can't get you free?" Phil posed.

"We'll call Graham Hawkes to find us with Wasp," I answered. Wasp, a one-atmosphere system that moves with thrusters, had been brought from England to serve as a backup.

We agreed that if we lost communication we would end the dive and return to the surface at once. During the previous week we had descended to 1,050 feet, then 740 feet; both times we had lost voice contact and had come up without touching the seafloor.

This time, the communications system worked faultlessly.

We slowly cruised for half an hour over gradually deepening terrain until we reached the sub's depth limit. Al finally announced, "Boh thinks we have to stop here. We're making inroads into the sub's power and air, and the current is picking up. We have to choose between a walk here and a much shorter walk later on."

We decided to stop amid more of the strange-looking spirals of bamboo coral, and near a broad, pale frond of soft coral so crowded with white polyps that it resembled a pear tree in full flower. We were at 1,250 feet. Air was vented from *Star II*'s ballast tanks as Boh stabilized the sub and made ready for my departure from the platform. Later Al admitted, "Turning the handle to release your waist strap and let you go was the most difficult move I've made in 10,000 dives. Phil, Graham, and I had been over the dozen things that could go wrong. But I kept thinking, 'What did we forget? No one has done this before. She's about to step into the unknown, and I can't swim out and get her if something goes wrong.' "

Al's concern did not show when he asked in a normal-sounding voice, "Are you ready?"

"Anytime," I responded. I was *more* than ready for this moment, to walk among the unusual animals, to look and touch, to explore at will. As I stepped from the platform, I was aware that I was venturing onto terrain in some ways comparable to the surface of the moon. Not only do the slopes and craggy ridges of the deep seafloor resemble a moonscape, but they also are equally unexplored. And, until recently, they have been just as inaccessible.

There is a significant difference, however. Astronauts on the moon have been impressed with its bleak, stark barrenness. Michael Collins, who orbited above, observed, "I have seen the ultimate black of infinity in a stillness undisturbed by any living thing." In the extreme darkness of the deep sea, a small circle of light from *Star II* illuminated for the first time a dozen or so long-legged, bright-red galatheid crabs swaying on the branches of a pink sea fan; a small, sleek, dark brown lantern fish darting by with lights glistening along its sides; an orange fish and several plumelike sea pens clinging to the rocky bottom near the edge of visibility. And when I turned away from the sub's lights, I could see sparks of living light, blue-green flashes of small transparent creatures brushing against my faceplate.

I longed to spring up and glide around, to follow a silky-skinned gray eel as it moved sinuously into a patch of light. Instead, I nodded sympathetically to a pale, joint-legged crab sidling along the ocean bottom. I felt as secure—and as bound to the seafloor—within my articulated metal suit as he was in his calcareous exoskeleton. Inside Jim, I walked and moved in slow motion, mastering the problems of weight and balance with a bearlike gait. Despite the lack of agility, I was mindful that at this depth, with 600 pounds of water pressing on every inch of my metal body, if I were breathing compressed gases I would have to spend at least a week in a decompression chamber before I could safely return to the surface.

As I wandered through the area, the sub powering along behind, I concentrated on observing the corals, especially the bioluminescent spirals of bamboo. Why do they pulse with light? Why do they glow at

DIVER MAGAZINE, LONDON

Armor for the sea's crushing depths, some early diving suits resemble gear now worn by astronauts. In a 1930 photograph British inventor Joseph S. Peress shows off "Iron Man" (above), a predecessor of Jim. The 800-pound suit enabled divers to reach depths of 400 feet—with a sea-level atmosphere inside. In tests during the 1920s, engineers lower "Deep Sea Steel Diver" (opposite) into Baltic waters. Like Iron Man, Steel Diver possessed articulated joints as well as claws for grasping. Such suits, designed for salvage operations, led to modern diving armor for industrial use and undersea discovery.

all? How do they and their neighbors survive in the eternal night of the deep sea? I still had a thousand unanswered questions when Al advised, "You have been out there two and a half hours."

"You're kidding," I responded. "It seems like 20 minutes." We had maintained a running dialogue, and my notebook was filled with observations. But Al was right. The time had sped away.

Aware of what I had just accomplished, with support from fine people and fine machines, I glanced at the two flags I had brought to the seafloor: one of the United States, the other, a sister flag to the one the National Geographic Society had sent to the moon. As *Star II* and I surfaced, I thought of the spirit of exploration they symbolized.

GRAHAM HAWKES

Like a fish on a line, the submersible Mantis *rises from the sea. Now used by the oil industry to inspect rigs and make repairs,* Mantis *can descend as far as 2,000 feet. Free from the peril of decompression and the need to breathe mixed gases, the operator uses manipulators and interchangeable "hands" to perform underwater tasks.*

In the deep sea we still have far to go, but creative minds are devising ways to extend our capabilities. More-flexible one-atmosphere diving systems are being developed. Sam, for example, is a smaller, lighter version of Jim that has more ringlike articulations and can be adapted for deeper excursions. And Graham Hawkes, designer of Wasp, has developed a one-atmosphere vehicle equipped with manipulators and several interchangeable "hands" for numerous industrial applications. This seagoing system, *Mantis,* is a one-person submersible powered by 10 propellers; some observers have likened it to an underwater flying carpet with robot arms. Says Hawkes, "It has worked in the Thistle [oil] Field, diving in 20-foot seas and 30-knot winds, inspecting the innermost structure at 600 feet. The other two units, *Mantis 2* and *3,* have completed check-out sea dives to 1,000 feet, prior to dispatch to the Arctic."

Hawkes, Nuytten, and other creative engineers all can imagine equipment—finely articulated glass or titanium suits—that may enable us someday to step into the ocean and dive freely to any depth. Meanwhile, others are probing the limits that the body can reasonably tolerate while exposed to the open sea.

SOME OF THE EXTREMES are encountered at the top of the earth—the North Pole. In all the world there is no place more hostile, less likely to tempt a would-be diver to take the plunge. But in early 1979 Joe MacInnis, undersea and under-ice pioneer, huddled with Al Giddings and NATIONAL GEOGRAPHIC Editor Gilbert Grosvenor in a hut at the Canadian Research Base of the Lomonosov Ridge Expedition (LOREX), drifting in the Arctic Ocean near the North Pole. For two months Canadian scientists had been studying the unfamiliar terrain below the floating ice. Sixty-four men in rotating teams conducted experiments and gathered information about the seafloor. Supported by geophysical and geological probes, the projects included photographing the bottom and its creatures, as well as measuring the deep-ocean currents and obtaining layerings from the ocean bottom. MacInnis and the others had accepted the scientists' hospitality. Now six feet of ice separated the trio from near-freezing water.

"Since we began working in the high Arctic in 1970, we've had over 750 dives, and on about 20 percent of them the regulators have failed. They freeze in the open position," said MacInnis. "Within minutes your tank is empty. Or sometimes it freezes shut. You inhale; your air isn't there. Here, every breath is a gift."

MacInnis, Giddings, and Grosvenor are sitting on a drifting, unsteady ice floe. Outside, the temperature is –31°F (–35°C). The wind blows and cuts through the tents. The ice groans as it cracks and forms fissures. For six months the sun does not set; it describes a circle above the horizon.

"The nearest land is under us—through 13,000 feet of seawater," MacInnis remarks. "Next nearest is about 450 miles away, at Ellesmere Island."

These three men are among the few to step through a hole in the ice into the incredibly clear sea below, to view the underside of the topside of the world. According to MacInnis, with the sea at 29°F (–2°C), below the freezing point of fresh water, "one of the greatest problems we had here initially was seeing. Moisture from our own breath, being more or less fresh, instantly freezes into a plaque of paper-thin ice on the inside of the mask. You have to take seawater into your mask, swirl it around, and clear it off. When your face is exposed to that cold water, it goes numb right away. The muscles don't hold the mouthpiece as strongly as they do in warmer water.

"The worst that can happen is to get a little water into the regulator, into the mouthpiece. When you inhale suddenly, cold water can go to the back of the throat and cause it to go into a spasm. You can't breathe. It's like inhaling a little jet of flame."

MacInnis then described another hazard: getting a large tear in one's protective dry suit. "There is a Niagara of freezing water around your skin. You have seconds to respond. It would be like falling into boiling acid—that's how painful it is under the ice. You must get out in a hurry."

But how?

Ringed seals under Arctic ice use the claws on their foreflippers to keep their diving holes open. Humans also must keep track of their entrance and go back the way they came. At the North Pole divers are covered by the nearly five million square miles of ice in the Arctic Ocean. From a hundred feet away, underwater, the single tiny window they carve to enter the sea resembles a postage stamp of blue—very small, very far away.

Despite the seemingly insurmountable difficulties, those who have dived under the ice at the North Pole are stunned by the incredible beauty of Arctic waters. Gilbert Grosvenor wrote of his dive: "The 24-hour sunlight trapped in the cobalt-blue fissures in the ice resembled lightning bolts frozen in motion. A crystalline chandelier—the condensation of fresh water—amazed me with its delicate etching."

Perhaps even more amazing is how easily such dives are achieved in this decade, compared with the difficulties of simply getting to the Pole a few generations ago.

Comdr. Robert Peary was the first to succeed, early in the century. Twenty years of trying, four excruciatingly difficult expeditions, brought him to the forbidding area where MacInnis, Giddings, and Grosvenor now sat. He finally reached the Pole because, as MacInnis put it, "Peary changed as a human being: He began to think like an Eskimo. He wore their furs, traveled the way they did; he used their dogs, their sleds, their food, and he became like them, integrated himself with the environment—and survived. To survive under the ice, we too had to adapt, to change our thinking and develop new technology for diving under extreme conditions.

"Walking upside down under the ice," said MacInnis, "I sometimes wondered if we weren't walking upside down *under* the footsteps of Peary. It has to be one of the most awesome experiences possible."

Exposure to another extreme is one focus of the work at the Duke University Medical Center, where a team of scientists and physicians headed by Dr. Peter Bennett studies the medical and physiological problems of pressure. Giant pressure chambers can simulate

U. S. NAVY

Deep-sea troubleshooter, a Cable-controlled Underwater Recovery Vehicle retrieves a torpedo off the coast of California. Designed to find ordnance fired during fleet exercises, CURV *helped recover a lost hydrogen bomb from the Mediterranean in 1966 and now performs jobs ranging from salvage and repair to biological sampling. In 1973 a* CURV *attached a line to a stranded* Pisces III, *enabling rescuers to raise that submersible and its two-man crew.*

conditions from 155,000 feet in altitude to more than 3,500 feet underwater, where pressure is 100 times that on the surface.

Bennett, an Englishman who came to Duke after 20 years with the Royal Navy's Physiological Laboratory, has a special interest in the "high pressure nervous syndrome"—a complex of reactions that plague divers beginning at about 500 feet.

CHUCK NICKLIN / SEA FILMS, INC.

In a dress rehearsal for a dive off the coast of Oahu, author Sylvia Earle, assisted by film producer Al Giddings, reaches out to close the dome-shaped hatch of Jim. Hoses from the rebreather lying atop the suit supply oxygen inside, where sea-level pressure of one atmosphere remains constant—even 2,000 feet down. Opposite, Dr. Earle beams from the open hatch. Gauges alongside her will measure depth and oxygen supply.

"They get tremors, dizziness, and nausea, and become fatigued and sleepy," Bennett told me. "Adding a small amount of nitrogen to the breathing mix of oxygen and helium helps to ameliorate the syndrome, but there is still much that we don't know about it. Its cause is one of the riddles that must be solved to make deep diving safe."

I asked Bennett what maximum depth a human could tolerate, and he flashed the boyish grin that is a Bennett characteristic. Obviously, it is a question he often asks himself.

"Not too long ago 300 feet seemed the impossible barrier," he said. "Use of helium has brought us to a maximum of perhaps 2,000 feet or a little deeper. But who knows? A new concept may give us free access to the greatest depths. Some whales can go a mile. Maybe we can, too, before long."

At the University of Pennsylvania, Dr. Christian Lambertsen, who years ago invented the oxygen and mixed-gas lungs used by the United States during and after World War II, is also optimistic. In chamber tests Lambertsen has studied physiological reactions of men to the pressure equivalent of 1,600 feet and found them able to work with reasonable effectiveness. And he has found that divers can breathe even when the density of the gas is equivalent to that which would be experienced breathing helium at a depth of 5,000 feet.

What are the limits, then? In 1975 Lambertsen wrote, ". . . oil producers have been stimulated to explore for (and produce) oil at depths to 1500 feet . . . with hopes of reaching 3000 feet within a few years and 6000 feet by 1980. In a single step of awakened interest, industrial and naval leaders have accepted the physiological feasibility of diving effectively to depths of at least 1000 feet of seawater. . . ."

So far, human beings have not been able to go even half a mile beneath the surface without the protection of a submersible. However, a number of years ago two other concepts that might extend the range were the subject of experiments. One, liquid breathing, floods the lungs with a superoxygenated liquid; the other surgically attaches artificial, membranous gills.

Dr. Johannes A. Kylstra, one of the handful of research pioneers in these areas, points out that many of the problems encountered in diving could be avoided if a noncompressible breathing mixture—a liquid—were used. In some of Kylstra's experiments adult white mice have lived as long as 18 hours immersed in a balanced, hyperoxygenated salt solution at 68°F (20°C). Despite such promising results, Kylstra concedes that liquid breathing by divers is still far in the future, particularly until the problem of carbon dioxide elimination is solved.

Frank J. Falejczyk worked with Dr. Kylstra for several years as the subject of some of these experiments. I asked him why he was willing to be subjected to the unknowns of liquid breathing. He replied, "I've been a diver for a long time and I share Dr. Kylstra's interest in learning about what the human body can do. Using liquid breathing for deep diving may or may not be possible, but the experiments we did already have had several useful carryovers for general medical practice."

Geologist Dr. Robert Ballard favors what seems to him to be a more efficient approach to underwater exploration. He puts the emphasis on discovery, not on the human condition. "Why expose hearts and lungs and livers to the dangers of going deep in the sea, when it is possible to take the most important part—the mind—there anyway?" Ballard asked after his recent deep dives in the Galapagos Rift. Using both the three-man submersible *Alvin* and the Acoustically Navigated Geological Underwater Surveyor (ANGUS), Ballard and his colleagues have had dramatic success in obtaining data from this rift at depths to 9,000 feet.

"Without ANGUS to first scout and document miles of terrain, it's likely that we would never have located the warm-water vents that have proven to be so interesting," Ballard continued. "Certainly, we saved an enormous amount of time and money traveling by using ANGUS to find the few small areas that merited a closer view with *Alvin.*"

Ballard, an enthusiastic scuba diver who has inspected more ocean bottom through *Alvin*'s ports than any other scientist, concedes that machines will never fully replace man in the sea.

"Nor should they," he continued. "There is no completely satisfactory substitute as yet for the combined dexterity, perception, and decision-making capability of an on-the-spot human being. But for general observation, as well as for well-defined specific jobs, machines are at least as good, sometimes better."

GERI MURPHY

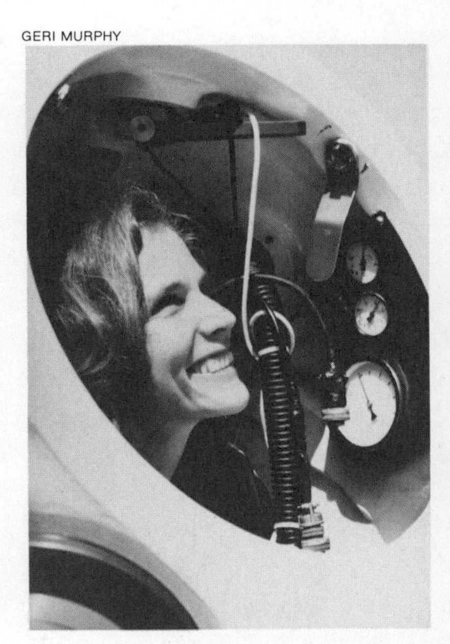

MECHANICAL EYES AND HANDS in the sea belong to a new generation of unmanned deep-sea vehicles. One of the most dramatic uses for one of these ingenious robots, called CURV (for Cable-controlled Underwater Recovery Vehicle), came during the tense weeks of March and April 1966. A hydrogen bomb had been lost in deep water near Spain, following an in-flight collision of two aircraft. After weeks of search by *Alvin,* the Perry *Cubmarine,* and *Aluminaut, Alvin*'s 10th dive paid off. The bomb was found in 2,500 feet of water, resting on a steep muddy slope, its attached parachute billowing above it. But raising the bomb proved almost as difficult as finding it.

During recovery attempts by *Alvin* and *Aluminaut,* including the first rendezvous of two submersibles in the deep sea, the bomb slipped away when a lifting cable broke. The bomb was lost and found three times before CURV was sent down to secure lines to the parachute. After days of painstaking effort by those on the surface, manipulating CURV's claw by remote control, two cables were attached. Then the current swept the nylon parachute over the vehicle's engines, hopelessly immobilizing the little machine.

The next move was to try to work CURV free by reeling in the attached lines. In a breath-stopping event, CURV, the chute, and the bomb all began to rise. Reported writer Robert Burgess: ". . . the lines kept coming a foot at a time, the long tedious, nerve-wracking pull lasting an hour and forty-five minutes. And then it surfaced, the CURV first, its tanks, tubes, and mechanical claw looking oddly out of place atop the billowing white canopy on which it perched."

On a visit to Edwin Link's Harbor Branch Foundation, I saw for myself some advantages of these remote-controlled vehicles. Aboard the aluminum-hulled *Sea Guardian,* I sat with engineer Fay Feild at the controls of CORD, a remote-controlled observation-and-rescue vehicle designed under the guidance of Link. Watching the television screen and flipping a control stick forward, back, or sideways to direct its movements, I soon felt that I—not the camera—was perched on the front end of CORD.

"It's not quite as good as getting wet myself," I told Feild, "but it is an excellent way to get a general idea of what's down there." I also knew CORD could do much that I could not do at any depth. It has a mechanical claw as well as a high-speed saw capable of cutting through steel cable, and it can be operated from *Sea Guardian.* Meanwhile, a television camera provides both immediate and permanent documentation on videotape.

CHUCK NICKLIN / SEA FILMS, INC.

Man-in-a-barrel hangs suspended undersea in a recreation of an early armored diving suit. Developed by Englishman John Lethbridge in 1715, the suit enabled him to salvage coins from shipwrecks. Lethbridge's design echoes in the Wasp, constructed and manned by engineer Graham Hawkes (opposite). In the thruster-powered diving system, Hawkes in 1978 went to a simulated depth of 2,000 feet in a British test chamber.

Some deep-sea Remote Controlled Vehicles (RCV's) as sophisticated as space probes are being used routinely to monitor and record certain kinds of work on offshore oil rigs. They sometimes go where the risk to a human diver would be too great.

Drew Michel, manager of Technical Services Division for one of the largest diving companies in the world, Taylor Diving and Salvage, described using an RCV in place of a diver. "In the process of monitoring the Ixtoc I well blowout in Mexican waters and helping with the installation of the 'sombrero' to contain the flow, we lost two of our vehicles. One we recovered, but the other has never been found. They are expensive—but replaceable. We could not send a diver where we sent those RCV's—our Harveys, as we have nicknamed them!"

The Harveys are not generally subjected to such extreme conditions, but they and other devices have found a number of appropriate uses in industry: from tagging along after divers as safety backups to performing certain placement and construction operations. One of these, TROV (Tethered Remotely Operated Vehicle), has worked in the Mediterranean and skied along the Arctic seabed while investigating the geology of the floor beneath those frozen seas.

Just as a wise mechanic does not discard his pliers simply because he has learned to use a wrench, those who enter the sea will undoubtedly use whatever is most appropriate for the purpose at hand, welcoming any tool that enhances their access to the sea.

IN SAN FRANCISCO a vision is taking form that should make it possible to open a full box of underwater tools, from mask, flippers, and snorkel to Aqua-Lungs, from hand-held cameras to remote-controlled "eyes beneath the sea." A 221-foot former U. S. Coast Guard Reserve training ship, the *Eagle,* is being converted for underwater research and documentary photography. She now sits at anchor in San Francisco harbor, but in the early 1980s she will set forth on a five-year expedition in the Pacific. I was with Al Giddings the first time he saw *Eagle* and spent hours with him, pointing a flashlight through the inner passageways, trying to imagine how a 100-man vessel designed for military training could be turned into a proper research vessel for 30, including crew.

Now gleaming blue and white, a diving bell and decompression chamber gracing the afterdeck, the *Eagle* is the focal point of Al's future. The next 25 years, he feels, will prove crucial for man's reliance on and use of the sea, and—through *Eagle*—he hopes to help transform indifference into respect and protection for our ocean resources.

His view is shared by Jacques-Yves Cousteau, who has said: ". . . the life cycle and the water cycle are inseparable, we must save the oceans if we want to save mankind." Pressing the limits of our ingenuity to explore and know the sea, understanding how natural systems work, and, most important, achieving harmony with the rest of the living creatures on this blue-and-white planet—these are among the most significant quests of our time. The outcome of civilization may well depend on how these challenges are met.

Like a magic carpet gliding through a liquid sky, the LRT *descends into Pacific waters carrying Sylvia Earle, inside the Jim suit, and the submersible* Star II *(below). Moments earlier, foaming seas had swirled around Jim (left) as the transport dipped beneath the*

CHUCK NICKLIN / SEA FILMS, INC. (ABOVE); SYLVIA A. EARLE (BELOW, LEFT); PETER ROMANO / SEA FILMS, INC.

surface; lights attached to Star II *protrude behind* Jim. *"I could hear the gurgling sounds of the sea," says Dr. Earle, "but it was frustrating to be inside and dry when I wanted to be wet and free to touch and feel the ocean." In a view from one of* Jim*'s four ports (opposite, below), Al Giddings prepares to photograph Dr. Earle during a test dive.*
FOLLOWING PAGES: *At 100 feet Dr. Earle practices moving about; a communication line entwines an 18-foot tether between* Jim *and* Star II. *"The* Jim *suit is very much like the astronauts' life-support systems," says Dr. Earle. "But I was an aquanaut on a walk through inner space."*

CHUCK NICKLIN / SEA FILMS, INC. (FOLLOWING PAGES)

100 AT
9

TERRY KERBY

"The light was faint, as on a night with no moon," says Sylvia Earle, peering through Jim's ports (left). "But when my eyes adjusted, the world I saw was incredibly beautiful." On the Pacific seafloor 1,250 feet down, she clutches a strand of luminescent bamboo coral (below). "For two and a half hours I wandered through fields of coral," she recalls. "The bamboo corals spiraled up like whiskers from the ocean floor, and when I touched them, rings of light pulsed up and down between base and tip. One sea fan was puffed with pink polyps like a fruit tree in blossom; red crabs dangled from it by their claws, waving in the current like shirts hung out to dry." Her walk completed, she ascends (right) and shows a piece of bamboo coral to Star II pilot Bohdan Bartko (above). Looking ahead to future deep dives, Dr. Earle says, "I'm ready!"

AL GIDDINGS / SEA FILMS, INC. (BELOW AND OPPOSITE)

CHUCK NICKLIN / SEA FILMS, INC.

N.G.S. PHOTOGRAPHER EMORY KRISTOF

Where no man dared to dream of diving 20 years ago, NATIONAL GEOGRAPHIC Editor Gilbert M. Grosvenor joins Al Giddings and diver-physician Joseph MacInnis in April 1979 for a descent beneath the ice near the North Pole. At the Canadian Research Base LOREX (above), the MacInnis team prepares for a chilling dive. With an electric auger and a shovel, the crew punches a hole in the ice floe (far left). Adjusting his mouthpiece, Al Giddings prepares to slip into 29°F (–2°C) waters (left); seated behind him, Gil Grosvenor awaits his turn. Popping up for a surprise appearance, a ringed seal peeks from the dive hole—and accepts an offering of mackerel. FOLLOWING PAGES: Under six feet of Arctic ice, Dr. MacInnis strolls upside down.

AL GIDDINGS / SEA FILMS, INC. (FOLLOWING PAGES)

To dive like a whale and breathe like a fish: Scientists search for ways to increase man's freedom and range in the sea. In a tank at a General Electric research laboratory (right), fish swim past a parakeet separated from their watery world by a transparent membrane. Watertight but porous, the membrane allows oxygen and carbon dioxide to pass in and out. Equipped with artificial gills, man may one day breathe liquid while diving. Inside a pressure chamber at Duke University Medical Center, Al Giddings films researchers (below) as they decompress from a four-day stay at a simulated depth of 1,500 feet. To study the effects of pressure on heart rate they prepare to hold their breaths (bottom, left), then immerse their faces in water (bottom, right).

GENERAL ELECTRIC COMPANY (RIGHT)

STEVE BOWERMAN / SEA FILMS, INC.

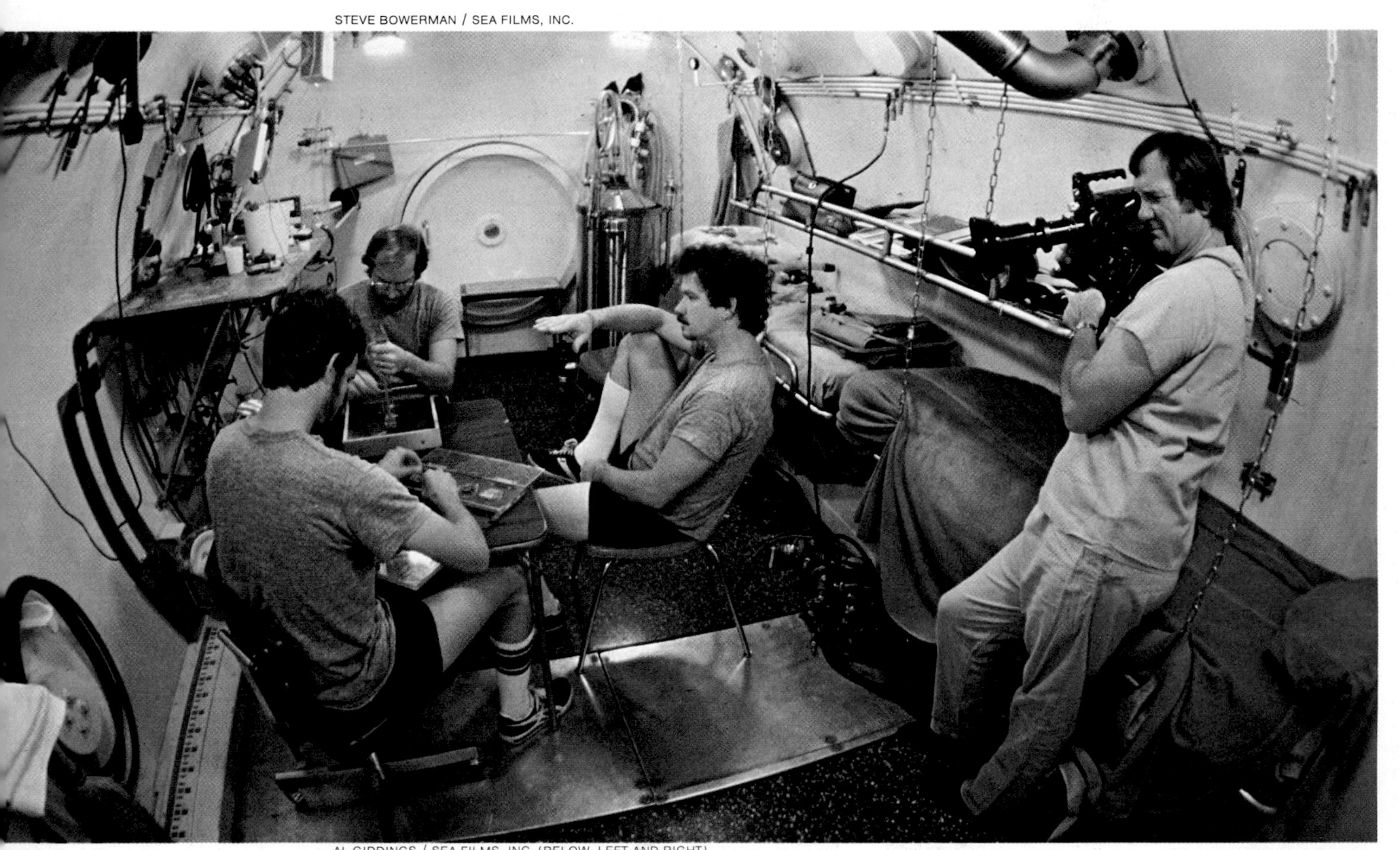

AL GIDDINGS / SEA FILMS, INC. (BELOW, LEFT AND RIGHT)

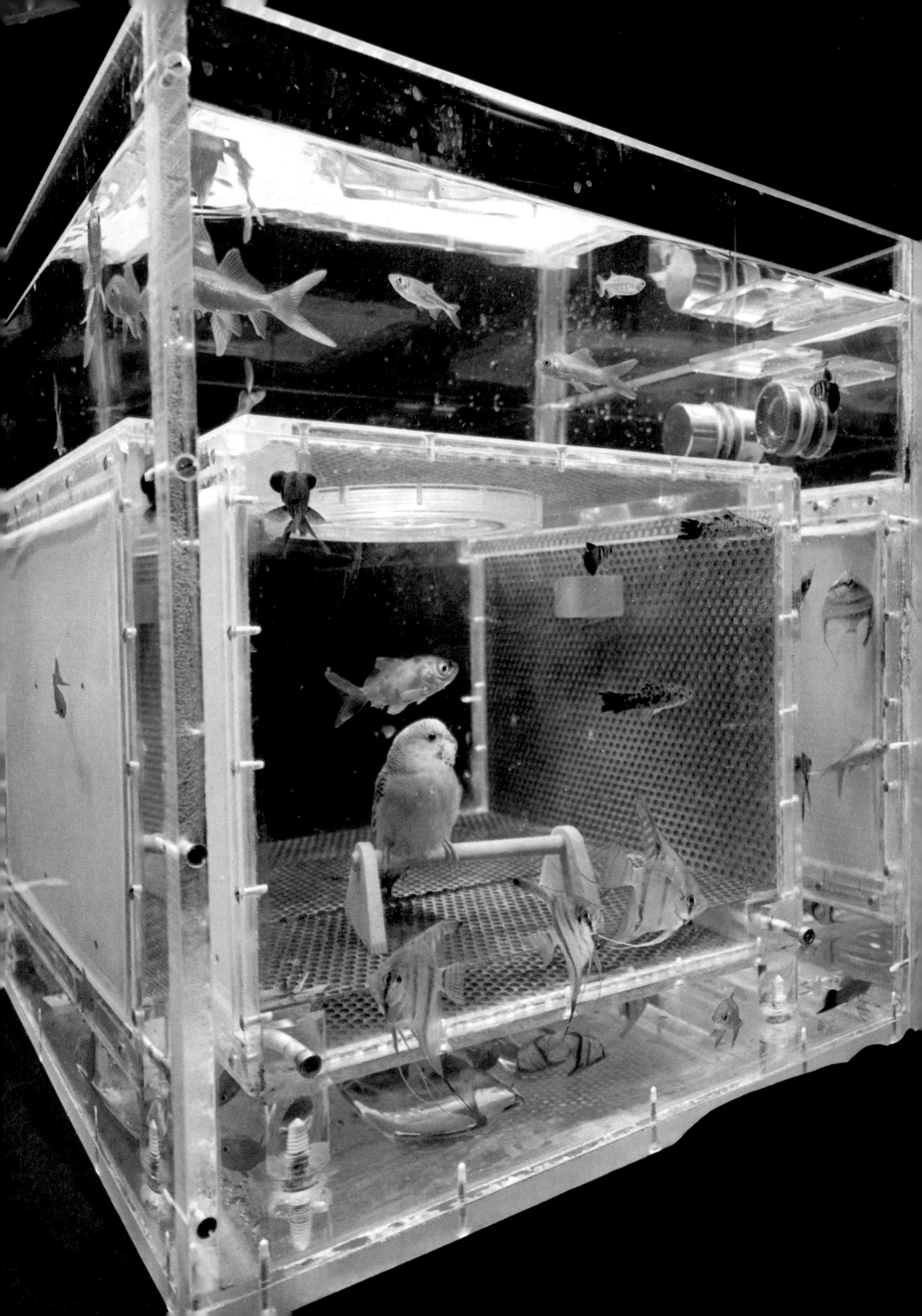

DREW MICHEL, TAYLOR DIVING AND SALVAGE CO. (ABOVE); U.S. NAVY (BELOW)

Eyes and hands beneath the sea, unmanned submersibles accompany divers and also probe waters too deep—or too hazardous—for divers to reach. During sea trials off the coast of California, technicians aboard the vessel Egabrag *(left) haul up "Harvey"—a Remote Controlled Vehicle, or* RCV. *Able to descend to 6,600 feet,* RCV*'s find wide use in the offshore oil industry, inspecting platforms and pipelines and surveying the seabed. In Pacific waters 200 feet deep, a narrowtooth shark (below) noses up for a closer look at the "flying eyeball." A painted face decorates the buoyancy tube of* Snoopy *(below, far left), a "swimming"*

JAN KOCIAN, HYDRO PRODUCTS

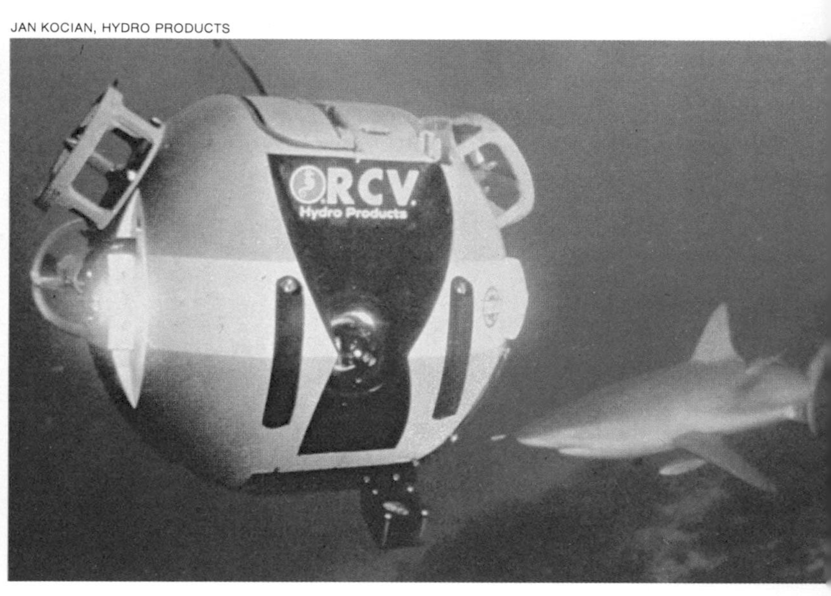

CHUCK COLLINS, SOLUS OCEAN SYSTEMS

television system deployed for shallow-water observation. In the Mediterranean off Algeria TROV *(below, left) works in tandem with a diver, providing light as he takes seafloor samples.*

WALTER CLAYTON, III / SEA FILMS, INC.

Dreams of a world-ranging ship for oceanographic study come true: A crew of workers aboard the research vessel Eagle converts the former Coast Guard Reserve training ship into a seagoing laboratory. On the afterdeck the team installs decompression chambers (above)—part of a saturation-diving system that will permit excursions as deep as 500 feet. Balanced on the yardarm 90 feet up, Rick Reinbold (opposite, top) prepares the mast for repainting; below deck, Jordan Coonrad (opposite, center) adjusts an underwater camera in one of the Eagle's numerous shops—which include a photographic lab, a sound room, an editing room, reference libraries, and aquariums. Owner Al Giddings envisions the vessel as "a standard-bearer for responsible ocean conservation policy" and plans to use the ship's facilities to produce television films on underwater subjects. At left, crouched amid an array of underwater equipment, Giddings speaks to a local television crew about Eagle's future as a staging platform for undersea expeditions.

Ready for a voyage of discovery, the research vessel Eagle cuts through calm waters in San Francisco Bay as a helicopter whirs toward her upper deck for a trial landing; Rick Reinbold directs the aircraft safely aboard (below). In the spirit of the H.M.S. Challenger's pathfinding expedition of a century ago, the Eagle crew plans to embark on a five-year journey at sea, sweeping in a great arc across the Pacific and stopping at a dozen locations for film production and marine research. Throughout history man has plumbed little of the watery realm that floods nearly three-quarters of the world; increasingly he looks to the sea for food and minerals, for treasure and a glimpse of the past—and to satisfy the human need to explore. "In our lifetimes we will face the most important decisions about the future of life in the sea," says Sylvia Earle. "Will we choose to keep it an Eden or witness a paradise lost?"

WALTER CLAYTON, III / SEA FILMS, INC.; JORDAN COONRAD (RIGHT)

Epilogue: An Ocean Ethic

Free as birds in flight, two snorkelers glide effortlessly above a school of grunts near the Galapagos Islands.

"If there is magic on this planet, it is contained in water. . . . Its substance reaches everywhere; it touches the past and prepares the future; it moves under the poles and wanders thinly in the heights of air. It can assume forms of exquisite perfection in a snowflake, or strip the living to a single shining bone cast upon the sea." Loren Eiseley, *The Immense Journey*

DAVID DOUBILET

A FIRST-TIME VISITOR TO EARTH, curious about the character of the planet, might logically choose to explore the ocean before considering continents. The sea is easily earth's most conspicuous feature, and within its broad expanse and great depths lies the most complete record of life available anywhere. To dive in the ocean is to immerse yourself in living history—to see, touch, and know fragile, ephemeral beings whose ancestry spans hundreds of millions of years. Most are sustained by a solution of elements similar to the saline stream that courses through our veins, a reminder of the kinship we have with other life and with the sea itself.

"I wish I was a fish," said Wart, the young King Arthur, in T. H. White's *The Once and Future King.* Wizard Merlyn obliged, turning the boy into a slim, swift creature with fins and scales. For a while Wart felt the delicious pleasure of swimming among the water weeds and getting acquainted with resident snails, mussels, and fishes.

To envision the world as Wart saw it, White advises, "imagine another horizon of under water, spherical and practically upside

Wreathed in a swirl of clouds, the earth gleams blue in the blackness of space. Oceans that give our planet its color hold keys to the origin of life and to earth's future. With boundless curiosity man probes the sunless depths that drop from shallow coastal shelves (below), explores the oceans' hidden places (bottom, right), and even photographs the violent underside of a breaking wave.

NASA (LEFT); AL GIDDINGS / SEA FILMS, INC.

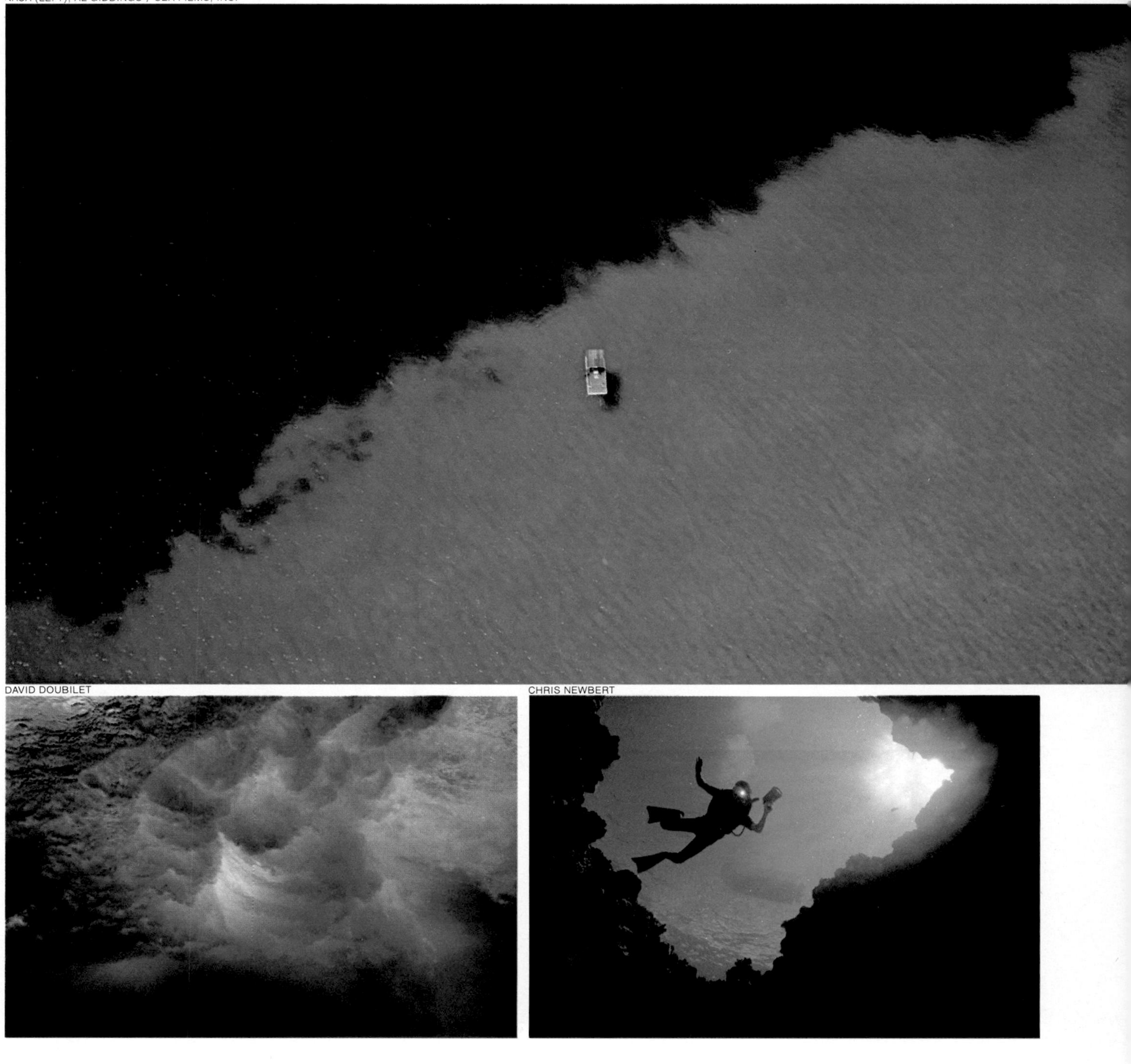

DAVID DOUBILET

CHRIS NEWBERT

JEFF ROTMAN

DAVID DOUBILET (ABOVE AND BELOW)

Reflection in a world of coral: Swimmers (right) glide placidly beneath a quicksilver surface that mirrors lush reef growth. A diver (above) perches momentarily on the edge of a coral mass in the Red Sea. At top, a school of grunts forms a flashing canopy over a Galapagos reef.

down—for the surface of the water acted partly as a mirror to what was below it. . . . Wart . . . was not earth-bound any more . . . pressed down by gravity and the weight of the atmosphere. He could do what men have always wanted to do, that is, fly. . . ."

To feel what Wart felt, to be lifted—weightless—from an earthbound view into another dimension, no magic is required. A mask and flippers will achieve the same effect for those willing to step into the sea, let go, and glide into the watery atmosphere that encompasses and gives substance to most living things.

We live in a time when scuba, submersibles, and other mechanical devices provide sophisticated access to the sea, and all have their places as we explore unknown frontiers. Microscopes and camera

lenses open another significant perspective, one that Merlyn automatically gave Wart when he made him small. The world, by and large, is populated by microbeasts. We are giants by comparison. A visitor to earth, closely inspecting life on this planet, might come to the conclusion that the most successful organisms here are less than six inches long and have no backbone. The invertebrates are overwhelmingly in the majority in terms of numbers and species, including hundreds of thousands of species other than insects.

Looking like plants touched by springtime, corals—actually animals—bloom as if by magic in an underwater garden. Lavender-eyed redlings (below) swim past a filigree of burning bush coral.

Even if we were half our present size, we would still dwarf most plants and animals. But with a lens we can take ourselves into the midst of the soft tentacles of an anemone, climb among coral castles side by

DAVID DOUBILET

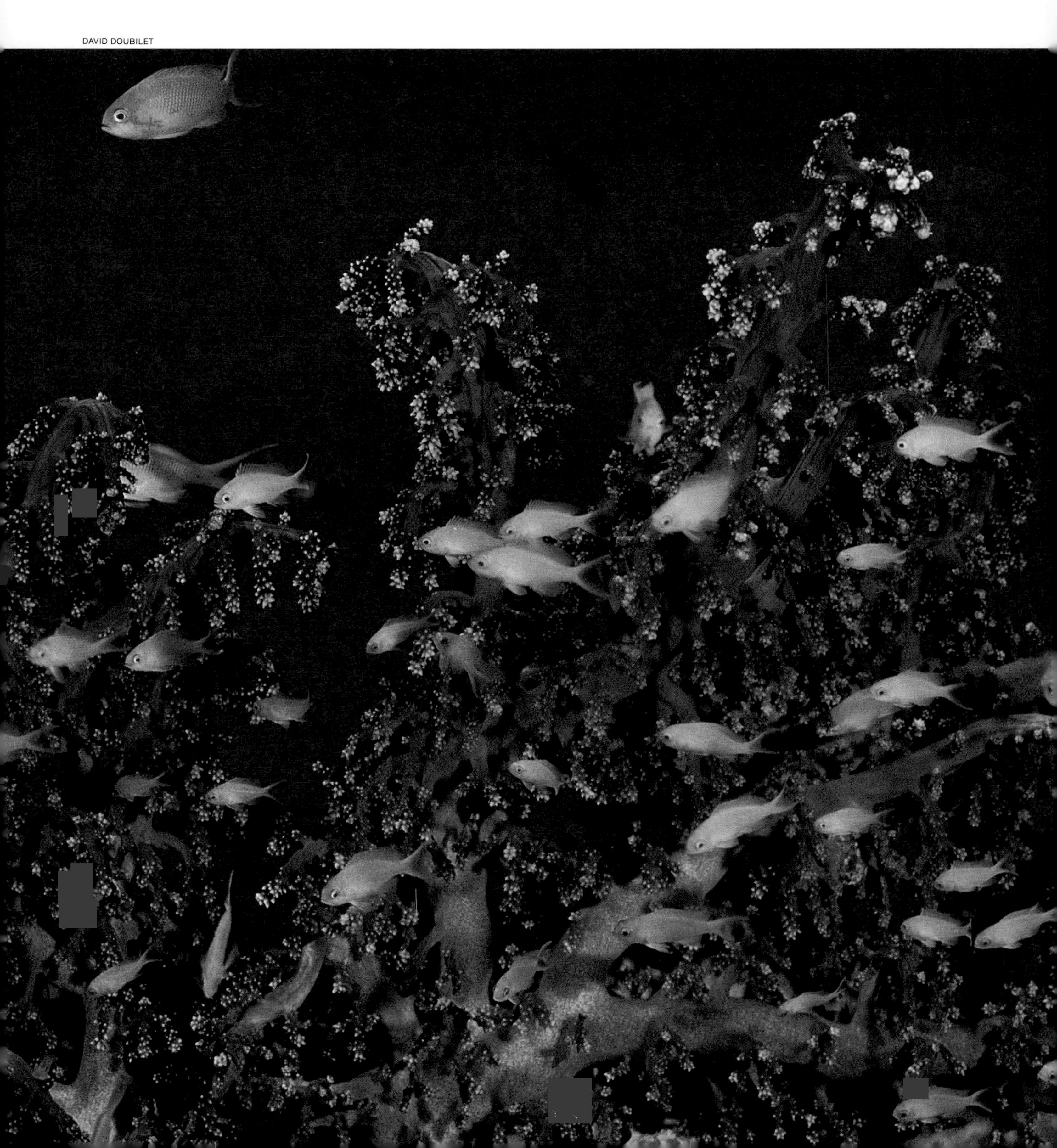

side with inch-long nudibranchs, explore the lives of starfish and sponges, and discover the inner workings of cells. Only close up is it possible to comprehend how dozens of strikingly different species can compatibly occupy a few square inches of reef space, or how a single kelp plant may support 100,000 microscopic animals.

Various technological creations have enabled us to discover that a few species are enormously abundant. There are trillions of certain kinds of bacteria, billions of small squid, millions of tuna. There are about 200 kinds of primates, including human beings, of whom there are now about four billion individuals.

Yet a species is not necessarily secure in its huge numbers. Only four

Tree coral (below) extends delicate polyps and gleans plankton from wafting currents.
A clownfish (bottom) nestles in the pink embrace of a writhing sea anemone.

CHRIS NEWBERT

JEFF ROTMAN

JACK MCKENNEY

Though clad in eye-catching reds, many marine animals live to blush unseen because water filters out the sun's red-reflecting rays.

DAVID DOUBILET (BELOW AND RIGHT)

JEFF ROTMAN

Bright coloration of the nudibranch (top), the branched corals (above and opposite), the gnarled sea raven (right), and the polka-dot starfish (FOLLOWING PAGES) *remains invisible to underwater man unless revealed by a diver's lamp or a photographer's flash.*

JEFF ROTMAN (FOLLOWING PAGES)

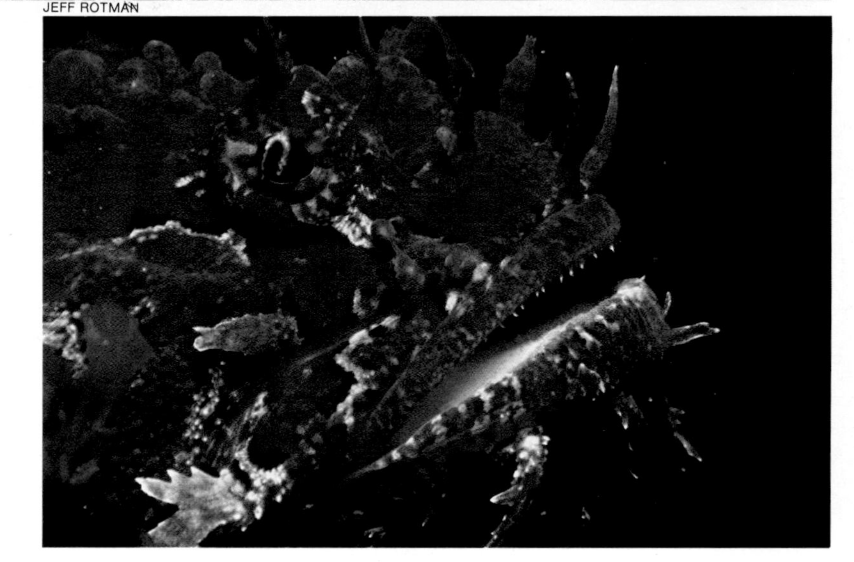

STEPHEN C. EARLEY (BELOW, LEFT AND RIGHT)

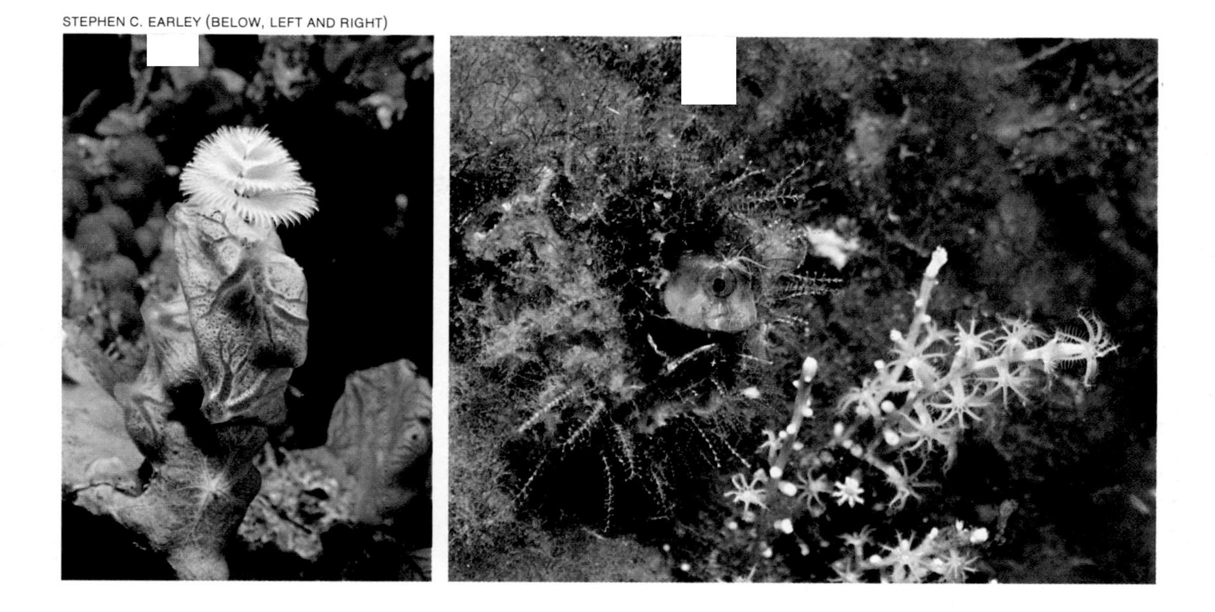

generations ago there were hundreds of millions of passenger pigeons in North America. Yet there are no passenger pigeons today—there have been none since 1914. The end for a highly successful group of animals may come quite suddenly.

Our destiny is linked to the fate of the oceans, and, ironically, we will determine the future of the sea through our actions in the next few decades. Either consciously, through intelligent choices, or by default, through ignorance or inaction, the present generation will have a magnified impact on the course of civilization. We can still

In an unending quest for food, surrealistic creatures of tropical waters creep or crawl, stalk, hide, or simply wait. A fragile arrow crab (opposite and below, left) picks its dainty way over an encrusting sponge. A tube worm (top, left) extends feathery parasols to catch passing plankton; inside a barnacle, a blenny finds shelter (top, right). Above, an azure sponge filters out current-borne food particles.

GERI MURPHY (LEFT AND BELOW)

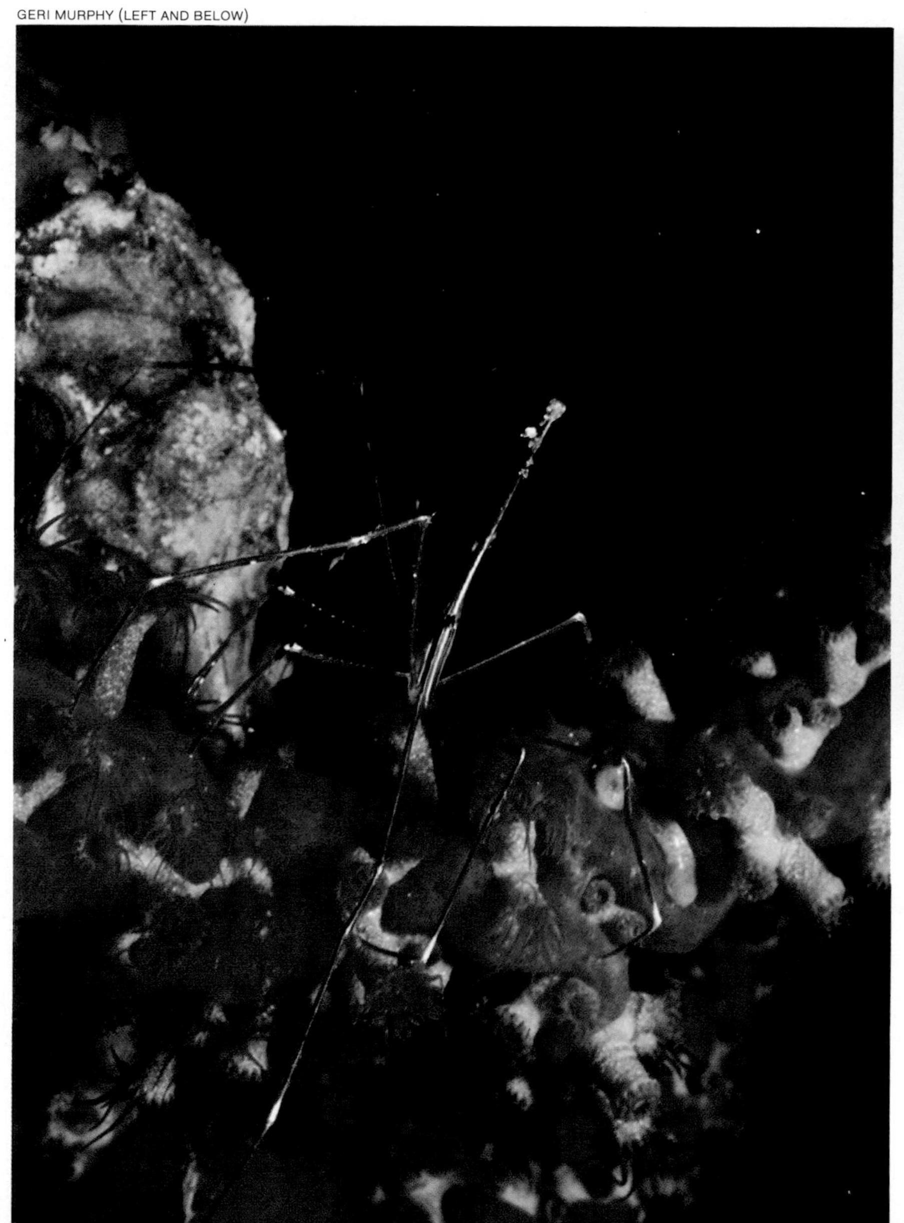

GERI MURPHY

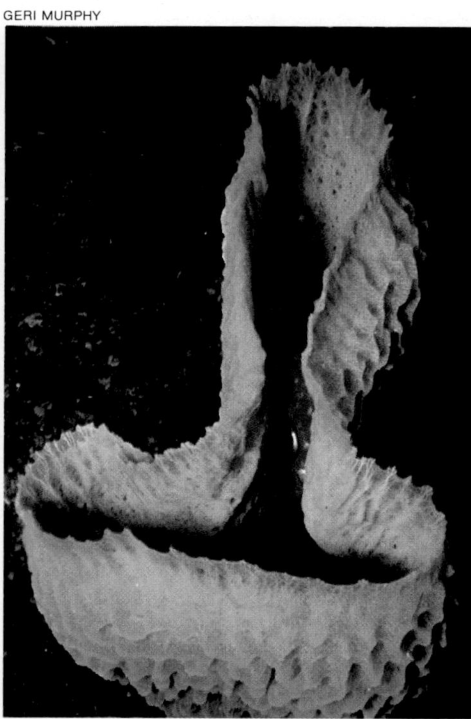

Fingerlike tentacles (below) form part of the nudibranch's digestive system; similar-looking limbs on tube-coral polyps (opposite, middle) and on a sea anemone (opposite, bottom) help these animals poison prey. A sea horse (opposite, top left) remains in place by grasping a sponge with its tail. Opposite, top right: A triton, one of the largest of the sea snails, extends striped appendages flanked by tiny eyes.

make choices concerning wilderness land and sea that will be denied another generation. In the past century we have already closed certain options, have changed the ocean more than during all preceding human history, and have set in motion events whose consequences future generations will inherit. Ignorance, not malice, has made the sea a receptacle for pesticides, other deadly chemicals, nuclear wastes. Ignorance, not malice, has caused us to overfish certain kinds of cod, herring, hake, and other sea species to the point where a large sustained take seems doubtful.

We still have much to learn about how natural systems work and how we can develop a harmonious, viable place for ourselves on the planet. Biologist Lewis Thomas remarked in 1979: "The only solid piece of scientific truth about which I feel totally confident is that we are profoundly ignorant about nature. . . . It is this sudden confrontation with the depth and scope of ignorance that represents the most significant contribution of twentieth-century science to the

DAVID R. LAUR

STEPHEN C. EARLEY

GERI MURPHY

JAMES M. KING

DAVID R. LAUR

DAVID DOUBILET (ABOVE AND BELOW)

Inquisitive to the point of nosiness, a sea lion nibbles on, then peeks into, a diver's camera housing. Some marine mammals, such as sea lions, remain tied to the land, where they mate, whelp, and bask. Others, like the Australian dugong mother and offspring at right—relatives of New World manatees—have become totally aquatic.

Never before has a generation been faced with such conscious decisions about the fate of the great whales or the mining of the deep seafloor—undisturbed for millennia. Never before have we had the opportunity to study such sea creatures as dolphins in close detail and to learn about—and from—them. Shall we continue to regard fish as simply a source of food? Or shall we look for other values that may cause us to protect the wildlife that naturalist Henry Beston calls "other nations, caught with ourselves in the net of life and time, fellow prisoners of the splendour and travail of the earth."

Many countries around the world are now grappling with the fate of one of the most numerous creatures on earth, a brilliant red species of krill, *Euphausia superba.* Between 800 million and five billion tons of these small, shrimplike creatures swarm in the nutrient-rich waters surrounding the least hospitable of all the seven continents—Antarctica. Disrupting the populations of krill may disturb an entire system that requires krill for food and includes whales, seals, seabirds, fishes, and squid—rather like nudging the cornerstone of a large building. Some experts estimate that we can safely double the protein removed from the sea by taking some of the

Glassy sweepers glow like a translucent tapestry in the Red Sea (below). Reefs there attract a yellow-head triggerfish (bottom, right), a showy creature that feeds on coral. Almost half a world away in the Caribbean, snails called flamingo tongues (below, right) feed on stalks of branched coral.

RAYMOND F. HIXON

GERI MURPHY

DAVID DOUBILET

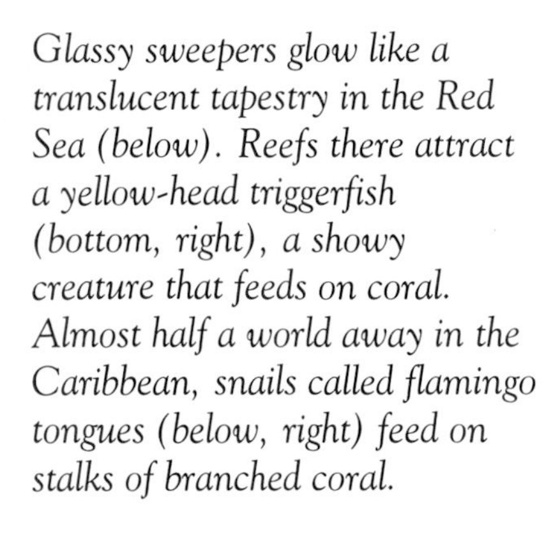

Flashing with purple, creole wrasses (right) mill around a domelike growth of little star coral in the warm, clear waters of the Caribbean.

STEPHEN C. EARLEY (RIGHT)

ourselves as a part of, not apart from, the tapestry of life on earth.

As a child I sometimes wished to be able to step into H. G. Wells's magnificent time machine, to go backward and walk softly along paths now covered with concrete. Or to leap ahead, to be able to whisk off to Mars for a Saturday afternoon, or to plan a meaningful expedition to a distant star. But our present era may be the most significant time in human history—a pivotal, decisive era.

Holding tight formation, spotted dolphins streak through Pacific waters. A familiar sight to mariners, the gregarious mammals often frolic alongside ships.

Since Marquis and his six-legged colleague mused about human nature, there has been the beginning of a shift that may be as significant as the Copernican revolution centuries ago, which determined, once and for all, that the earth was not the center of the universe. Astronauts looking back on earth and aquanauts looking upward from the sea have—after millions of years—changed the perspective of mankind. Perhaps for the first time we are seeing

CHRIS NEWBERT

Secure in its ponderous armor, a loggerhead turtle paddles serenely, undisturbed by a photographer's bubbly presence. An octopus (below), bulbous eyes narrowed to slits, retreats from a diver's unsettling light. As if focused on a central point, sting rays gather in a circle, perhaps engaging in some form of courtship behavior.

CHARLES ARNESON

BERNIE CAMPOLI (ABOVE); JACK MCKENNEY (BELOW)

human intellect. We are, at last, facing up to it . . . we are getting glimpses of how huge the questions are, and how far from being answered. . . . But we are making a beginning, and there ought to be some satisfaction, even exhilaration, in that."

How different is this attitude from that related by Archy, the literate cockroach created half a century ago by New York newspaperman Don Marquis. Archy encountered a self-satisfied toad, Warty Bliggens, who

"considers himself to be
the center of the said
universe
the earth exists
to grow toadstools for him
to sit under
the sun to give him light. . . ."

Says Archy,

"if i were a
human being i would
not laugh
too complacently
at poor warty bliggens
for similar
absurdities
have only too often
lodged in the crinkles
of the human cerebrum."

BEN CROPP

krill; others feel that removing *any* piece of what may be a precariously balanced system not only may jeopardize whales and penguins but also may affect the stability of the entire southern sea.

The time has come to combine the wisdom of science and the sensitivity of art to mold attitudes that will transcend written laws. Needed is an ocean ethic. Traditionally the sea has been regarded as a common heritage for all mankind; now its care must be acknowledged as a common responsibility. It is in the best interests of all of us to develop international policies that recognize the interdependence of life and the need for nations together to maintain the basic elements of life support. A century ago the ocean was wilderness. Before we are gone, we may choose to keep what remains of that sea of Eden, or cause—and witness—paradise lost.

DAVID DOUBILET

Glinting silver in a haze of blue, barracudas (opposite), spadefish (above), and big eyes swim in tight-packed schools. FOLLOWING PAGES: *Sweepers mass in a cascade of sunlit gold.*

DAVID DOUBILET (FOLLOWING PAGES)

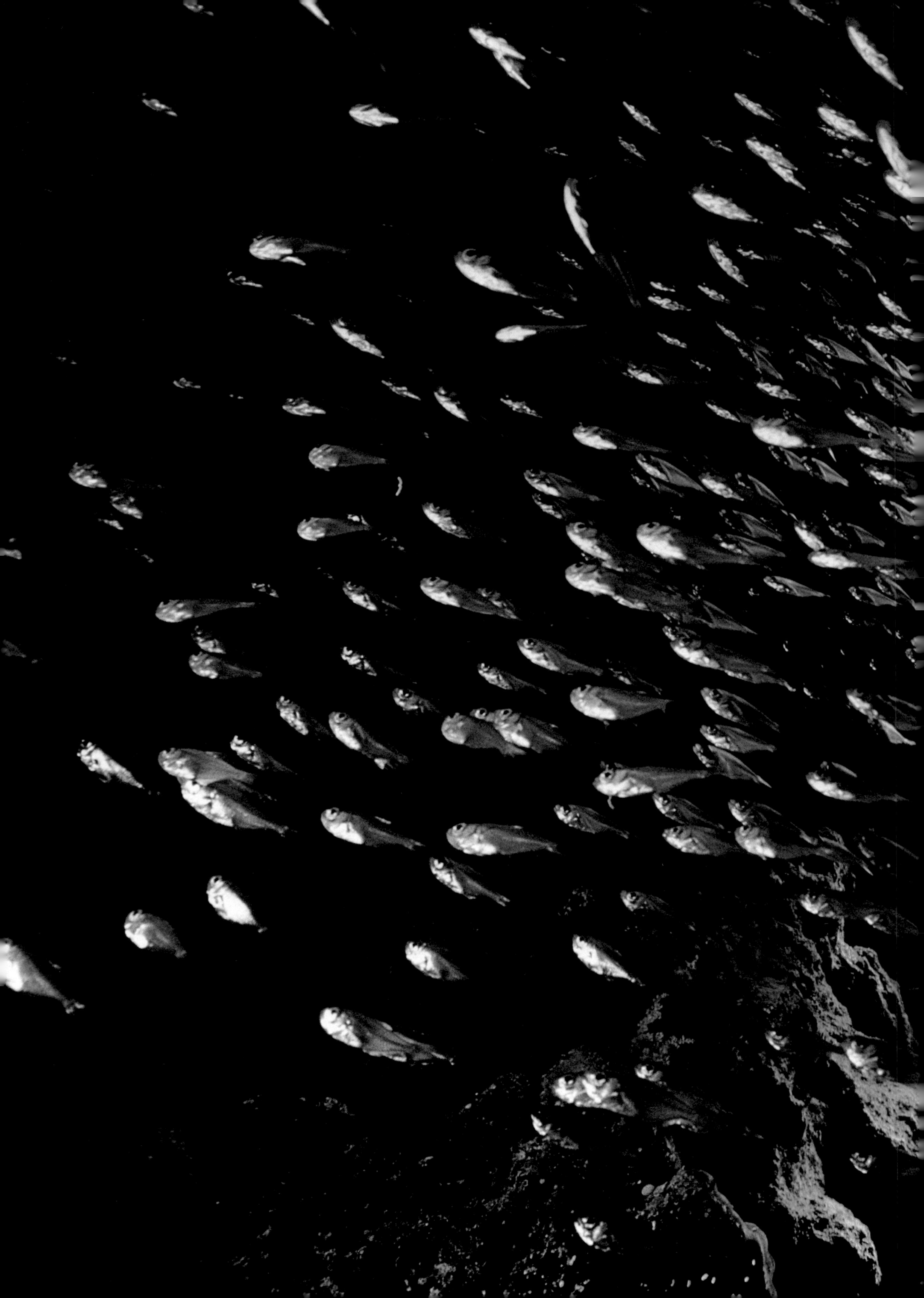

Additional Reading

The reader may wish to consult the *National Geographic Index* for related articles and to see the Society's publications *The Ocean Realm* and *Undersea Treasures.* The following books may also be helpful:

Bern Anderson, *By Sea and By River: The Naval History of the Civil War*
John E. Bardach, John H. Ryther, and William O. McLarney,
Aquaculture: The Farming and Husbandry of Freshwater and Marine Organisms
George F. Bass, *Archaeology Beneath the Sea*
William Beebe, *Half Mile Down*
P. B. Bennett and D. H. Elliott, eds., *The Physiology and Medicine of Diving and Compressed Air Work*
Henry Beston, *The Outermost House*
Mark M. Boatner, *The Civil War Dictionary*
Elisabeth Mann Borgese and Norton Ginsburg, eds., *Ocean Yearbook I*
Robert F. Burgess, *Ships Beneath the Sea: A History of Subs and Submersibles*
R. Frank Busby, *Manned Submersibles*
Rachel L. Carson, *The Sea Around Us*
Bryan Cooper and T. F. Gaskell, *The Adventure of North Sea Oil*
Jacques-Yves Cousteau, *The Living Sea* and *The Silent World*
Robert H. Davis, *Deep Diving and Submarine Operations*
Margaret Deacon, *Scientists and the Sea, 1650-1900: A Study of Marine Science*
James Dugan, *Man Explores the Sea: The Story of Undersea Exploration From Earliest Times to Commandant Cousteau* and *Man Under the Sea*
Nicholas C. Flemming, ed., *The Undersea*
John T. Foster, *The Sea Miners*
Perry W. Gilbert, ed., *Sharks and Survival*
C. P. Idyll, *The Sea Against Hunger* and (ed.) *Exploring the Ocean World: A History of Oceanography*
Howard E. Larson, *A History of Self-Contained Diving and Underwater Swimming*
Peter R. Limburg and James B. Sweeney, *Vessels For Underwater Exploration*
Marion Clayton Link, *Windows in the Sea*
Eric Linklater, *The Voyage of the Challenger*
Frank Lipscomb, *Historic Submarines*
Robert F. Marx, *Into the Deep* and *Shipwrecks of the Western Hemisphere*
Richard K. Morris, *John Holland 1841-1914: Inventor of the Modern Submarine*
National Oceanic and Atmospheric Administration, *The NOAA Diving Manual: Diving for Science and Technology*
Navy Department, *U. S. Navy Diving Manual*
Kenneth S. Norris, ed., *Whales, Dolphins, and Porpoises*
Walter Penzias and M. W. Goodman, *Man Beneath the Sea: A Review of Underwater Ocean Engineering*
Jacques Piccard, *The Sun Beneath the Sea* and, with Robert S. Dietz, *Seven Miles Down*
Carl Proujan, *Secrets of the Sea*
David A. Ross, *Opportunities and Uses of the Ocean*
C. W. Rush, W. C. Chambliss, and H. J. Gimpel, *The Complete Book of Submarines*
Susan Schlee, *The Edge of an Unfamiliar World: A History of Oceanography*
Terry Shannon and Charles Payzant, *Windows in the Sea*
Edward H. Shenton, *Diving for Science: The Story of the Deep Submersible*
Charles Shilling, Margaret Werts, and Nancy Schandelmeier, eds., *The Underwater Handbook: A Guide to Physiology and Performance for the Engineer*
Brian J. Skinner and Karl K. Turekian, *Man and the Ocean*
E. J. Slijper, *Whales*
James B. Sweeney, *A Pictorial History of Oceanographic Submersibles*
John Sparks and Tony Soper, *Penguins*
Lewis Thomas, *The Medusa and the Snail*
Charles Thomson, *The Voyage of the Challenger,* 2 vols.
James Vahan and James Dugan, eds., *Men Under Water*
Jules Verne, *Twenty Thousand Leagues Under the Sea*
Richard C. Vetter, ed., *Oceanography, the Last Frontier*

Hovering diver scatters a group of tiny redlings above a sea fan on a Red Sea reef. A wealth of marine life brings scientists and divers to the incredible coral formations of this region, home of many species of plants and animals found nowhere else on earth.

DAVID DOUBILET

Highlights of Diving History

The story of undersea exploration is a chronicle of drama, danger, and discovery. The charts and the chronology on these pages highlight a few of man's more memorable adventures in the sea. Some of the symbols stand for one-of-a-kind exploits, such as *Trieste*'s 35,800-foot plunge to the bottom of the Mariana Trench. Others were chosen to represent the kind of scientific research carried out in a certain era. For example, only one undersea habitat is pictured—Jacques Cousteau's Conshelf Three—though many other habitats proved equally important.

The illustration above reveals the continental shelf—that area sloping from the shoreline to a depth of 600 feet. The drawing on the following pages presents the abyss, from the edge of the continental shelf to the deepest known place in the sea. To fit the abyss on a single page, it was necessary to compress the part of the scale that indicates depths below 5,000 feet. To identify each event illustrated, look for its symbol in the historical listing.

Fourth century B.C. Early written record of a device for supplying air to divers: Aristotle describes Greek sponge divers breathing air trapped in kettles lowered into the water.

Alexander the Great's diving chamber: Persistent legend tells of the Macedonian king's descent in a "barrel" or "cage" of glass.

First century B.C. 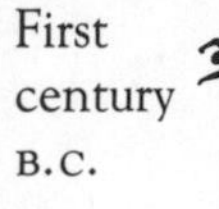 Diving women of the northwestern Pacific gather food from the seafloor during breath-hold descents. *Ama* of Japan continue the tradition today.

1535 Guglielmo de Lorena designs a bell that divers use to explore sunken barges in Lake Nemi in Italy.

1578 Englishman William Bourne describes a primitive submarine in *Inventions and Devices.*

1620 Dutch inventor Cornelius van Drebbel stages tests of one of the first workable submarines, in the Thames River in England.

1663 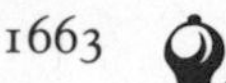 Using a diving bell, salvagers recover cannon from the sunken warship *Vasa,* lying 110 feet down in Stockholm harbor.

1680 Giovanni Borelli of Italy describes a self-contained leather diving device.

1690 In England Edmond Halley designs a diving bell that uses weighted barrels lowered from the surface to renew the air supply. Divers later use it to descend more than 50 feet.

1715 Englishman John Lethbridge develops one of the first armored diving dresses—a wooden cylinder with a glass view port and laced leather sleeves—and uses it 60 feet below.

1776  First attack by a military submarine: David Bushnell's *Turtle* tries to sink H.M.S. *Eagle* in New York harbor.

1797 K. H. Klingert of Germany invents a helmet diving dress supplied with pumped air.

1800 Robert Fulton of the United States tests his submarine, *Nautilus,* in France.

1819 Appearance of a practical open-dress diving suit fed with pumped air, in England: The invention is generally credited to Augustus Siebe.

1837 Siebe designs an airtight closed-dress diving suit vented by a valve.

1839 Divers wearing Siebe diving dress begin salvage of the sunken warship *Royal George* off Spithead, England. The British Royal Navy founds the first diving school at the wreck site.

1844 Henri Milne-Edwards of France and a naturalist friend conduct the first underwater studies of marine life, in the Strait of Messina, off Sicily.

1850 Wilhelm Bauer's submarine *Brandtaucher* causes Danish blockaders to retreat from Kiel harbor, Germany. Bauer's crew performs the first submarine escape from 60 feet in 1851.

1863 Simeon Bourgeois and Charles-Marie Brun build *Le Plongeur,* a submarine run by compressed air, in France.

1864 Confederate submarine *Hunley* sinks the U.S.S. *Housatonic* in an attack near Charleston, South Carolina, and is itself destroyed in the battle.

1865 Frenchmen Benoît Rouquayrol and Lt. Auguste Denayrouze develop a breathing apparatus with an automatic demand valve and a reservoir of air mounted on the diver's back.

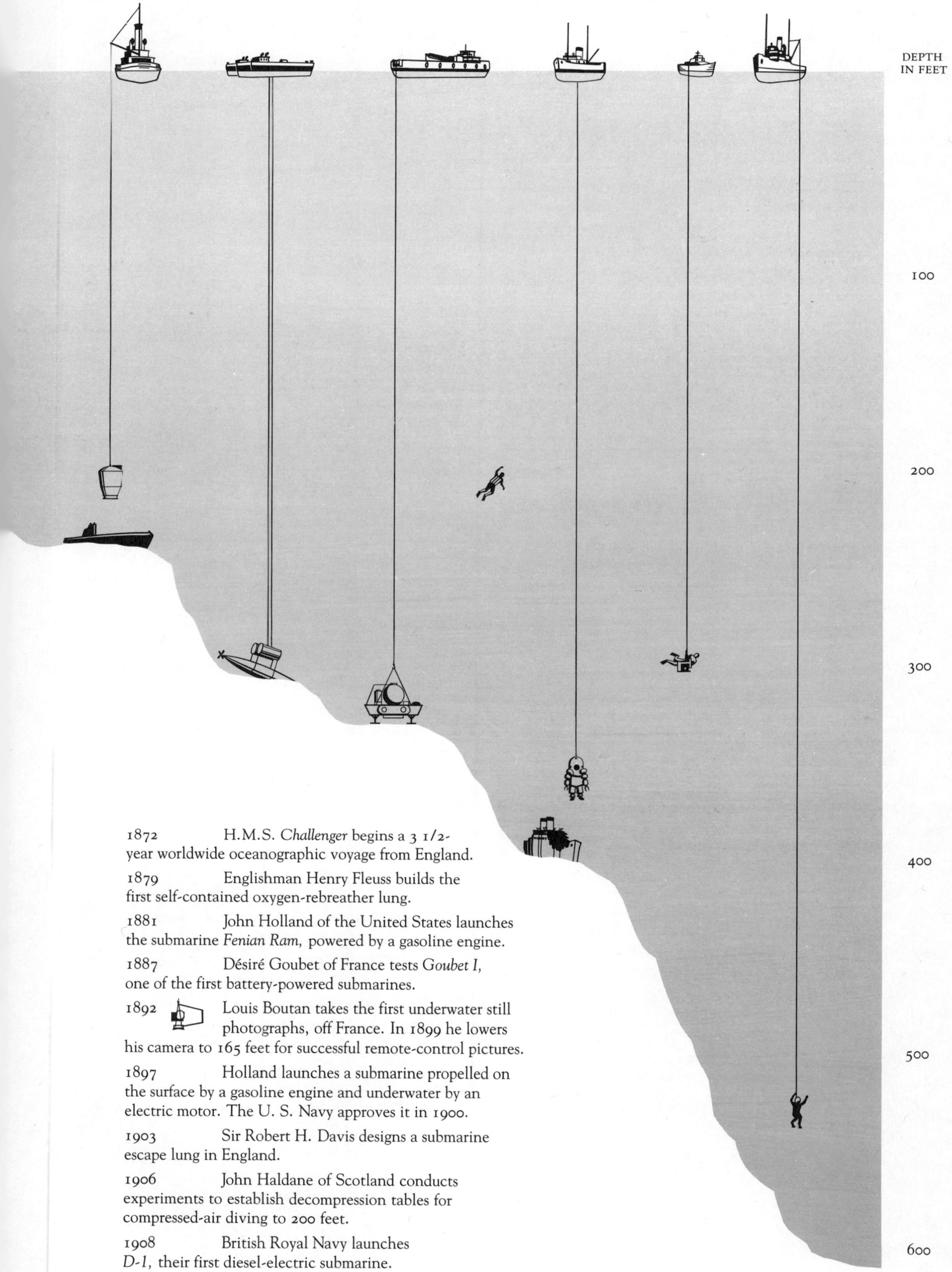

DEPTH
IN FEET

100

200

300

400

500

600

1872 H.M.S. *Challenger* begins a 3 1/2-year worldwide oceanographic voyage from England.

1879 Englishman Henry Fleuss builds the first self-contained oxygen-rebreather lung.

1881 John Holland of the United States launches the submarine *Fenian Ram,* powered by a gasoline engine.

1887 Désiré Goubet of France tests *Goubet I,* one of the first battery-powered submarines.

1892 Louis Boutan takes the first underwater still photographs, off France. In 1899 he lowers his camera to 165 feet for successful remote-control pictures.

1897 Holland launches a submarine propelled on the surface by a gasoline engine and underwater by an electric motor. The U. S. Navy approves it in 1900.

1903 Sir Robert H. Davis designs a submarine escape lung in England.

1906 John Haldane of Scotland conducts experiments to establish decompression tables for compressed-air diving to 200 feet.

1908 British Royal Navy launches *D-1,* their first diesel-electric submarine.

1912 Davis invents the first pressurized, submersible decompression chamber.

1913 Neufeldt and Kuhnke, a German manufacturing firm, patents an armored diving dress with ball-and-socket joints in the arms and legs.

1915  Depth record for recovery of an entire vessel: Salvagers raise the submarine *F-4* from 306 feet off Honolulu.

1917 English Capt. Guybon Damant leads efforts to recover gold worth five million pounds sterling from the sunken liner *Laurentic,* lying at 132 feet off North Ireland.

1924 Frenchman Yves Le Prieur invents a manually valved, self-contained compressed-air lung.

British inventor Joseph Peress builds an armored suit with oil in the joints of the articulated arms and legs.

1926 William H. Longley and NATIONAL GEOGRAPHIC's Charles Martin take the first artificially lighted underwater color photographs, off the Dry Tortugas, Florida.

1927 Helium undergoes initial tests as a diving gas at the Experimental Diving Unit in Washington, D. C., following pioneering experiments by J. H. Hildebrand, R. R. Sayers, and W. P. Yant.

1928 U. S. Navy adopts C. B. Momsen's submarine escape lung with an oxygen-rebreathing system.

1930 British Admiralty Deep Diving Committee establishes decompression tables for depths to 300 feet.

Using a bathysphere, naturalist Dr. William Beebe and designer Otis Barton dive to 1,428 feet off Bermuda.

Divers wearing Neufeldt and Kuhnke suits raise five tons of gold bullion from the *Egypt,* sunk in 426 feet of water off France.

1932 Beebe-Barton bathysphere dives to 2,200 feet off Bermuda.

1934 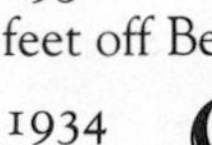 Beebe-Barton bathysphere reaches 3,028 feet off Bermuda.

1935 Jim Jarratt, wearing the "Iron Man" armored suit, locates the sunken *Lusitania,* 330 feet down off Ireland. Modern Jim suits are named for him.

Louis de Corlieu markets rubber foot fins in France.

1937 American engineer Max Nohl, breathing a helium-oxygen mix, dives to 420 feet in Lake Michigan.

Momsen and Lt. K. Wheland of the Experimental Diving Unit accomplish a simulated dive to 500 feet, breathing a helium-oxygen mix.

1939 *Squalus* sinks 243 feet in waters off Portsmouth, New Hampshire; the 33 survivors are brought up in the McCann rescue chamber. The *Squalus* is raised during the next four months, marking the first open-sea use of helium as a diving gas.

1943 Frenchmen Jacques-Yves Cousteau and Emile Gagnan perfect the fully automatic, compressed-air Aqua-Lung and use it to dive to 210 feet in the Mediterranean.

1945 Arne Zetterström demonstrates the use of hydrogen on a dive to 525 feet in the Baltic Sea. A mishap during decompression causes his death.

1948 British Royal Navy diver Wilfred Bollard, breathing a helium-oxygen mix, dives to 540 feet in Loch Fyne, Scotland.

Frédéric Dumas dives to 307 feet using an Aqua-Lung designed by his colleague Jacques Cousteau.

Auguste Piccard and Max Cosyns test the bathyscaph *F.N.R.S. 2* off Dakar, Senegal.

1954  Frenchmen Georges Houot and Pierre Willm pilot the bathyscaph *F.N.R.S. 3* to 13,287 feet off Dakar.

1955 First nuclear submarine: The U. S. Navy launches *Nautilus.*

1956 Swedish divers begin to raise *Vasa* from 110 feet in Stockholm harbor.

1960 U. S. nuclear submarine *Triton* circumnavigates the globe.

In the bathyscaph *Trieste,* Jacques Piccard and Lt. Don Walsh, USN, descend to 35,800 feet in the Mariana Trench, the deepest known place in the sea.

1961 Hannes Keller of Switzerland and writer Kenneth MacLeish, wearing wet suits, dive to 728 feet in Lake Maggiore, Switzerland.

1962 Keller and Peter Small use a diving bell to plunge 1,000 feet off Santa Catalina island; they breathe a special mix of gases developed by Keller. Small dies during decompression.

 French bathyscaph *Archimède* dives to 31,308 feet in the Kuril Trench off Japan.

Man-in-Sea, headed by Edwin A. Link: During this saturation diving project, Belgian Robert Sténuit descends in a decompression chamber to 200 feet off the French Riviera and stays 24 hours.

Continental Shelf Station (Conshelf) One, led by Cousteau: Frenchmen Albert Falco and Claude Wesly live for seven days in 33 feet of water off Marseille.

1963 Conshelf Two: In the Red Sea off Port Sudan five men spend one month at 36 feet, and two men stay seven days at 90 feet.

Trieste explores the site of the sunken nuclear-powered submarine *Thresher,* at 8,400 feet.

1964 Man-in-Sea: Robert Sténuit and Jon Lindbergh submerge to 432 feet in the Bahamas and stay two days.

Sealab I, organized by Capt. George F. Bond, USN: Four Navy divers stay 11 days at 193 feet off Bermuda.

1965  Sealab II: Twenty-eight men remain for 15 to 30 days at 205 feet in California waters.

Conshelf Three: Six men live 22 days at 328 feet in the Mediterranean.

1966 Submersibles *Alvin, Aluminaut,* and the Perry *Cubmarine,* search for a hydrogen bomb lost near Palomares, Spain, in more than 2,500 feet of water. The bomb is recovered by the unmanned vehicle CURV.

1967 Link builds *Deep Diver,* the first modern submersible with an operational lock-out chamber.

1968 U. S. Navy diver Robert Croft descends to 240 feet in a breath-hold dive off Bimini in the Bahamas.

1969 Tektite I, sponsored by the U. S. Navy, the U. S. Department of the Interior, and NASA: Four men live for 60 days in 50 feet of water off the U. S. Virgin Islands.

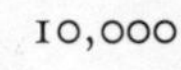

Submersible *Ben Franklin* drifts in the Gulf Stream for 1,500 miles, reaching depths of 2,000 feet on its 30-day journey. Jacques Piccard heads the expedition.

1970 Tektite II: Eleven successive five-person teams, including one all-woman team, spend from 14 to 20 days at 50 feet off the U. S. Virgin Islands.

Aegir, sponsored by NOAA: Six men remain submerged for six days in 520 feet of water off Hawaii.

1972 COMEX, an underwater service-and-construction company in France, carries out a simulated dive to 2,001 feet in a dry chamber.

1976 Deepest dive in a Jim suit: Oceaneering International's one-atmosphere system goes to 1,440 feet off Spain to recover a TV cable.

Frenchman Jacques Mayol dives to 328 feet in a breath-hold dive lasting 3 minutes 40 seconds off Elba.

1977 Deepest working saturation dive: COMEX divers go to 1,510 feet. They later accomplish a record dive to 1,644 feet.

1978 Research ship *Glomar Challenger* drills into seafloor sediment through waters 23,104 feet deep.

1979 Sylvia Earle, in a Jim suit, dives to 1,250 feet off Oahu, riding the submersible *Star II* to the seafloor.

1980 Depth of nuclear submarine: Official figure lists depth at more than 400 feet; others estimate more than 1,500 feet.

Divers at Duke University Medical Center go to a simulated depth of 2,132 feet, breathing a mix of helium, oxygen, and nitrogen.

ILLUSTRATIONS (ABOVE AND PRECEDING PAGES) BY MARK SEIDLER

N.G.S. PHOTOGRAPHER BATES LITTLEHALES

CLAUDE E. PETRONE, NATIONAL GEOGRAPHIC STAFF

In far-flung seas National Geographic photographic expertise contributes to the steady progress of underwater exploration. Above, Dr. Richard Chesher uses an OceanEye camera housing designed by photographer Bates Littlehales to record seafloor phenomena off the U. S. Virgin Islands. At left, photographer Emory Kristof (with dark beard) and design machinist Alvin M. Chandler make a final shutter adjustment before sending an acoustically controlled camera to explore the Galapagos Rift. Opposite, Kristof and Sam Raymond, an executive of Benthos, Inc., prepare to lower jointly developed cameras 13,000 feet into the Arctic Ocean through a hole in polar ice.

Acknowledgments

The Special Publications Division is grateful to the individuals, organizations, and agencies named or quoted in the text; to our chief consultant, Albert R. Behnke, Jr.; and to those cited here for their generous cooperation and assistance during the preparation of this book: Alice Alldredge, Arthur J. Bachrach, Ken Bassett, Frederick Bayer, Peter B. Bennett, Thomas E. Berghage, Hans Bertsch, Stefan Blasco, Robert C. Bornmann, Michael J. Borrow, Tracy Bowden, R. Frank Busby, Floyd Childress, James J. Childress, Stephen C. Earley, Lev Fishelson, Sarah Goodnight, Jim Greenslate, Royal Hagerty, Paul E. Hargraves, John Harrison, Trevor Haydon, F. G. Hochberg, Suk Ki Hong, Meredith L. Jones, Robert Kanazawa, Nancy Kaufman, Thomas LaPuzza, Joseph D. Libbey, Raymond Manning, Luis Marden, John E. McCosker, George McCourt, Thomas J. McIntyre, Daniel McShae, James Mead, Gary Morrison, John Mulligan, Katherine Muzik, John Newton, James Norris, Kenneth S. Norris, John Oshinski, Wallace Raabe, B. G. Rees, George Rees, Clyde Roper, Joseph Rosewater, Klaus Ruetzler, Richard H. Ryan, David Seefeldt, Gene Shinn, Carl I. Sisskind, Hans Soop, Richard Speer, Victor Springer, Leif Thornvall, Ruth D. Turner, Fred van Doorninck, J. R. Weber, John Wells, John P. Wise, George R. Zug; American Petroleum Institute, Deepsea Ventures, Inc., Institute of Nautical Archaeology, Monitor Research and Recovery Foundation, National Marine Fisheries Service, National Maritime Museum of Sweden, National Oceanic and Atmospheric Administration, Phillips Petroleum, Scripps Institution of Oceanography, Smithsonian Institution, Submarine Force Library and Museum, Tuna Research Foundation, Undersea Medical Society, Inc., U. S. Department of Interior, U. S. Geological Survey, U. S. Navy, and Woods Hole Oceanographic Institution.

Boldface indicates illustrations; *italic* refers to picture legends (captions)

The Special Publications Division wishes to thank the following copyright holders for permission to reprint material in the text:

Lines from "warty bliggens, the toad," which appeared in *archy and mehitabel* by Don Marquis, copyright 1927 by Doubleday & Company, Inc. Reprinted by permission of the publisher.

Lines from *The Bottom of the Sea* by A. M. Sullivan, copyright 1966 by Dun & Bradstreet. Reprinted by permission of the publisher.

Lines from "Little Gidding" in *Four Quartets* by T. S. Eliot, copyright 1943 by T. S. Eliot; copyright 1971 by Esme Valerie Eliot. Reprinted by permission of Harcourt Brace Jovanovich, Inc., and Faber and Faber Ltd.

HARDCOVER: DRAWING BY JODY BOLT FROM PHOTOGRAPHS BY N.G.S. PHOTOGRAPHER BATES LITTLEHALES AND DAVID DOUBILET.

N.G.S. PHOTOGRAPHER EMORY KRISTOF

NATIONAL GEOGRAPHIC SOCIETY

WASHINGTON, D. C.

Organized "for the increase and diffusion of geographic knowledge"

GILBERT HOVEY GROSVENOR
Editor, 1899-1954; President, 1920-1954
Chairman of the Board, 1954-1966

THE NATIONAL GEOGRAPHIC SOCIETY is chartered in Washington, D. C., in accordance with the laws of the United States, as a nonprofit scientific and educational organization. Since 1890 the Society has supported 1,817 explorations and research projects, adding immeasurably to man's knowledge of earth, sea, and sky. It diffuses this knowledge through its monthly journal, NATIONAL GEOGRAPHIC; its books, globes, atlases, filmstrips, and educational films; *National Geographic WORLD*, a magazine for children age 8 and older; information services; technical reports; exhibits in Explorers Hall; and television.

Library of Congress CIP Data
Earle, Sylvia A. 1935-
Exploring the deep frontier.
Bibliography: p.
Includes index.
1. Underwater exploration. 2. Diving, Submarine. I. Giddings, Al, 1937- joint author. II. National Geographic Society, Washington, D. C. Special Publications Division. III. Title
GC65.E185 910'.09162 80-7567
ISBN 0-87044-343-7

Composition for EXPLORING THE DEEP FRONTIER by National Geographic's Photographic Services, Carl M. Shrader, Chief; Lawrence F. Ludwig, Assistant Chief. Text sections printed by Federated Lithographers-Printers, Inc., Providence, R.I. Color sections printed by Holladay-Tyler Printing Corp., Rockville, Md. Bound by Holladay-Tyler Printing Corp., Rockville, Md. Color separations by The Beck Engraving Co., Philadelphia, Pa.; The Lanman Companies, Washington, D. C.; National Bickford Graphics, Inc., Providence, R.I.; Progressive Color Corp., Rockville, Md.; The J. Wm. Reed Co., Alexandria, Va.